COMING BACK TO
EARTH

Cornford tackles some of the biggest challenges facing our communities (climate, urbanisation, food systems) by taking us back to the essence of the message of the Gospel. If we believe that the good news of Jesus is about life-changing transformation and reconciliation, then our discipleship must be relevant in ways that engage with these challenges. This work provides a great balance of both deep theological reflection combined with ideas for practical application thus enabling the reader to consider what it means to be modern day 'salt and light'. The result is both prophetic and hope-filled. A must read for every Christian who is serious about the practical application of their faith in a world facing huge systemic challenges.

Matthew Maury,
National Director, TEAR Australia

Amidst the babble of stories that disconnect us from the earth and each other, Cornford's brilliant and passionate telling of the biblical story not only dares to name the sources of our own grief and pain, but insightfully pulls us in to the deeper, richer story of our own calling as the suffering followers of Jesus. The vision of this book is no pie-in-the-sky dream, rather it is a way of walking for those of us who want to root our households in down-to-earth, faithful, joyful, kingdom living, no matter what the consequences. If you are hungry for a fully embodied picture of how the Bible calls us to live in faithfulness with the earth and each other, then this book is for you.

Sylvia Keesmat,
biblical scholar, author of *Colossians Remixed*

Jonathan Cornford is one of Australia's most radical Christian thinkers. This short book is a pithy and profound summary of much of his recent writing. Its challenges to our western, over-stuffed, ecologically destructive lifestyles come thick and fast. But they are delivered with a lightness of touch, grace, humour, and deep pathos. Jonathan is a weeping prophet. He and his family live with these challenges daily, but not self-righteously. This book is down to earth, honest, even blunt, but delivered with healthy doses of biblical hope. It deserves, and our perilous plight demands, a large readership.

Revd Dr Gordon Preece,
Director, Ethos - EA Centre for Christianity & Society

Cornford has done a great job recontextualizing the Sabbath Economics vision into the Australian context, and maintains a keen commitment to justice, sustainability and the church. This collection of essays offers theological, pastoral and practical resources to the demands of discipleship in a world groaning under ecological and social crisis.

Ched Myers,
theologian, author of *Binding the Strongman*

COMING BACK TO
EARTH

Essays on the Church,
Climate Change,
Cities, Agriculture
and Eating

JONATHAN CORNFORD

a. Acorn
Press

Published by Acorn Press
An imprint of Bible Society Australia
ACN 148 058 306 | Charity licence 19 000 528
GPO Box 4161
Sydney NSW 2001
Australia
www.acornpress.net.au | www.biblesociety.org.au

ISBN 978-0-647-53417-5

First published by Morning Star Publishing in 2016,
ISBN 978-0-994-26455-8

NATIONAL LIBRARY OF AUSTRALIA A catalogue record for this work is available from the National Library of Australia

Cover and text design and layout by John Healy

CONTENTS

Acknowledgements

I would like here to acknowledge my deep debt and gratitude to Peter Chapman, whose teaching, mentoring and conversation, not to mention feedback on earlier drafts, have been so helpful in the shaping of the thoughts contained herein. Of course, any deficiency in such must accrue to myself. Similarly, I would like to thank Kim, my wife, who reads and corrects everything I write, who is always encouraging, and who is a pillar of the thinking and lived experience that has shaped this work. Lastly, I would like to thank Amy and Mhairi, my daughters, who keep everything clear. It is for them and the world they grow into that these thoughts are offered.

'So Shall We Reap' first appeared as an article in *Zadok Perspectives*, Autumn 2014.

'Shaking the House' first appeared as a *Zadok Paper* in Summer 2014.

Jonathan Cornford, All Saints Day 2014.

INTRODUCTION

Humanity at the beginning of the third millennium is faced with unprecedented challenges. The list falls on upon our ears with a mournful and dangerous familiarity: climate change; species extinction; resource depletion; pressure on the global food system; widening international tensions and conflicts; economic instability and fragility; persisting poverty and economic exploitation; vast and growing social inequality; global people movements on a scale not seen since the end of the Second World War. And then there are the peculiar maladies of 'affluenza' that have gripped the wealthy world. In one sense, none of these challenges are new; but in another sense the sheer scale and dimension of them is indeed novel.

Those of us in the wealthy Western world live in a bizarre juxtaposition between, on the one hand, our awareness of these profound challenges and unprecedented access to information about them; and on the other hand, our immersion in a culture that provides perpetual temptation into triviality and distraction. Many of us live with a massive cognitive dissonance of knowing that we face these challenges while living day to day as if they did not exist. Nowhere is this more evident than in the very urgent and direct challenges presented by climate change where the action we must take is fairly clear, and yet the scale of inaction has been monumental.

And for many Western Christians, there is a third element to this strange dissonance. On the one hand we believe – or say we do – that 'God so *loved* the world', that the life and teaching of Jesus is a '*light* to all peoples', that 'in Christ God was *reconciling* the world to himself', and that the whole thing is *good news* in the largest possible sense of the term; and yet on the other hand, there is little recognition amongst the mass of the membership of the church that Christianity has anything distinctive or important to say about the unprecedented challenges we are now faced with. Somehow Christianity is good news for the world and yet has nothing to say in its hour of need.

It is no wonder that the upcoming generation is turning away from the faith of their parents in droves. Whether their eyes are upon the attraction and temptations of consumer culture or upon the plight of the world, or both, Christian faith just seems so irrelevant to *life*.

This group of essays wells up from a deep sense of protest against this state of affairs. At its base is the affirmation that in Jesus, God's *Word* – what God has to say to humanity – has taken on human form, and that this Word is indeed the light of all people, in the fullest, broadest and deepest sense. Three of the essays each examine a foundational challenge that must be confronted this century – climate change, urban life and food – and each finds that the biblical narrative has something *essential* to say about each of them: essential in that *we need to hear it*; and essential in that it penetrates to the very *essence* of the problem and its solution. Whether it be the practicalities of decarbonising the economy, finding a human scale in the midst of monstrous cities, or conserving topsoil for agriculture, we can find in the Bible both prescient explanations for our current predicament and pertinent wisdom that offers hopeful guidance. The final essay seeks to explore in more depth some of the theological and practical complexities of seeking to live by the sorts of ethical frameworks that are called forth by such challenges. We need to re-learn a biblical way of asking ethical questions, and the ways in which Jesus and Paul deal with the ethics of eating is highly instructive.

William Temple once wrote that 'Christianity is the most avowedly materialistic of all the great religions'[1], by which he meant that the Christian gospel is indivisibly spiritual *and* material – the Word must *always* become flesh. For some Christians this may seem strange territory; it has been widely understood that Christianity is merely a spiritual message about getting to heaven and we have not been accustomed to turning to the Bible for solid guidance on the material

---

1 William Temple, *Nature, Man and God*, Macmillan, 1934, p.478.

and economic affairs of the world. But that misshapen belief is a theological and historical aberration and the first essay in this collection asks how it is that modern Christianity has diverged so significantly from the faith of its founder and Lord. The church in the West has come to a crossroads and it is no accident that this is at a time when humanity has come to a crossroads.

Central to all of these essays is the word 'crisis' and it pertains to both the state of global affairs and the state of the church. For some, such a word sounds a note too shrill and too sharp – it is discomfiting to rational and measured discussion. But as Abraham J. Heschel reminds us, the prophetic voice of the Bible is continually discounted as 'one octave too high':

> For us the moral state of society, for all its stains and spots, seems fair and trim; to the prophet it is dreadful.[2]

Our word 'crisis' stems from the Greek word *krisis*, which in our New Testaments is most frequently translated as 'judgment' and sometimes as 'justice'. That is, crisis is a time when things come to a head, when a crunch point is reached and there must be some sort of accounting. We have reached such a point with climate change, the church in the West is very rapidly coming to such a point, and our cities and food system may well be approaching such a point faster than we think.

However, we have generally misconstrued crises in the same way we have misconstrued judgment in the Bible – we see a crisis as an unpleasant end that is to be denied or avoided. However, most of the judgement proclaimed in the Bible – especially that of the prophets and of Jesus – is not *final* judgment, but rather an *opportunity* to change direction. It is not for nothing that the gospel writers record Jesus' proclamation as beginning with 'Repent'! But here again we have misconstrued a core Biblical concept: repentance (*metanoia* in Greek) does not merely refer to sorrow for one's sins but coming to a critical change in direction; it literally means to put on a new mind or a new

---

2 Abraham J. Hescehl, *The Prophets*, Vol. 1, Harper Colophon Books, 1962, p.9.

perception. Central to the message of the Bible – and of these essays – is the understanding that the sooner *we* can come to such a 'crisis', to true repentance, the sooner we can begin walking the path of life, which is the only place where true hope is to be found. Every crisis is indeed an opportunity in disguise.

It is for this reason that this collection of essays is also concerned to not leave discussion of 'global' challenges or of 'Christianity' as abstract issues, but as challenges that penetrate to the heart of *our lives* and *our faith*. In most of these essays, discussion of the great challenges of our times ends up in our homes. We are not responsible for changing the world but we are responsible for our part in it, and it is in the everyday economics of our lives that the issues of our times require a response.

The most pressing challenges of faith are always, firstly, to see the world clearly, and secondly, to ask what God requires of us in such a world. Faith, hope and love take on most substance when they are *enfleshed* in practical and visible ways; but finding the ways to do this in a complex global consumer economy is by no means straightforward. This modest collection of thoughts cannot claim to answer such questions, but it is hoped that it can offer some guidance in pursuing them.

STANDING AT THE CROSSROADS
Where the crisis of the church and the crisis of the world meet

Stand at the crossroads, and look, and ask for the ancient paths,
where the good way lies;
and walk in it, and find rest for your souls.
Jeremiah 6:16

If we are honest, many of us would have to admit to a deep ambivalence about church – about the church as an institution, and about the whole Sunday thing. Often this ambivalence is made up of a number of elements, and we are not always clear within ourselves about its sources. There is often discomfort or disappointment about both the history and present reality of the institution that claims to itself the gospel of Jesus. It is hard to overestimate just how deeply and how widely revelations such as those concerning child abuse and subsequent cover-up within the church have shaken people's ability to place some faith and hope in it as an institution.

At a much more mundane level, there is, I believe, a widespread disappointment and discomfort that the membership of the church so poorly reflects the character and work of Jesus of Nazareth. How often has someone's exciting discovery of the world-inverting implications of the gospel turned to bitterness in the face of the inertia and apathy of the local congregation? Belief for so many Christians has become an exercise in rationalising the cognitive dissonance between what they read in the Bible and what they encounter on Sunday mornings.

On the other hand, the church, like many other social institutions, is feeling the crunch of rapid cultural change. The core practice of the church – which is simply to gather – has been placed under enormous strain by our modern social geography and the virtual surrender of its members to consumer culture. The dislocation of church membership from geographic locality has transformed the experience of 'belonging' to a church community into yet another act of consumer choice.

Moreover, the Sunday service has, for many, become an exercise in frustration and disappointment: some are exasperated with the stultifying strangeness of traditional worship; others are disturbed by the consumer predilections of 'contemporary worship'; many are uneasy with the superficiality of the whole thing and its irrelevance to the great challenges of humanity in our age: climate change, refugees, family breakdown, mental health, poverty and conflict.

Underpinning all of this is a much deeper malaise: Christianity in the West is undergoing something of a crisis of faith. As we shall see, it is not just that many mainline churches are undergoing a sustained numerical decline that jeopardises their future viability – that presents challenges enough – but that the *faith* of so many still in the pews, and even of those in leadership, has become so uncertain. Just what is the good news of Jesus and what does it mean for how we live in the world?

This is a sombre assessment of the church today. How did we get here and what must we now do? These two questions cannot be asked enough, and the answers must be sought by looking both deep and wide. This essay will explore one strand of this question – a critical strand to be sure, but by no means the whole picture – by exploring the relationship of Christianity to our material lives. What relationship does the spiritual basis of our being have to the physical dimension of our existence? Might it be that the deep uncertainty and ambivalence of contemporary Christianity is rooted in its profound disconnection from the things that matter?

The economics of the Body of Christ

To gain a clearer picture of the present position of the church it is helpful to ask ourselves again what the church *is meant to be*. The Greek word in the New Testament that is translated as 'church' – *ekklesia* – simply means 'the gathering'. It was a word common in the Greco-Roman world and it was usually used in reference to gatherings related to civic governance –the *ekklesia* denotes a body politic. When the Apostle Paul uses the word 'church', he is very often using the term in its most ordinary sense, referring to specific gatherings of Christians in particular locations – the churches in Galatia or Macedonia or Judea. However, sometimes (particularly in Ephesians and Colossians) when Paul speaks of 'the church', he is referring to something much more than a local gathering; for example: 'He has put all things under [Christ's] feet, and made him, as he is above all things, the head of the Church' (Eph 1:22, NJB). Paul doesn't anywhere make entirely clear what he means by this higher concept of 'the church' and interpretations differ significantly along Protestant, Catholic and Orthodox lines. Nevertheless, whatever the interpretation, it is clear that Paul understands followers of Jesus to participate in an essential unity that is far wider and far deeper than any particular gathering they may be associated with.

Paul's understanding of both the essence and the unity of the church is most profoundly captured by his description of it as 'the body of Christ'. When Paul invokes this term, he is not, as is so often presumed, merely using it as a metaphor, nor is he dressing-up an institution in religious language, and neither is he conceiving of some abstract, mystical reality that we can never quite see. Rather, he is saying something much more startling, and much more sobering in its implications. When Paul refers to the body of Christ, he is saying the same thing as the Gospel of John when it tells us that 'the Word became flesh' (Jn 1:14).

The great scandal of Christianity is that God's communication with humanity – the Word (*logos*) – took the form of a human life, that of a carpenter's son in an obscure corner of the Roman Empire. The Word became flesh because what God wants to say to us concerns flesh. But Paul takes this idea further. He asserts that, following the resurrection and ascension of Christ, and through the sending the Holy Spirit – indeed, 'the Spirit of Christ' (Rom 8:9) – the church is now *the body* of Christ; it is indeed the continuing incarnation of Jesus in the world.

The truth of the gospel is that the Word must always become flesh. When we read the rest of the Bible in this light, we see that the process of Word becoming flesh is actually God's modus operandi the whole way through – it is the meaning of Israel, the charism of the prophets, the revelation that is in Jesus, and it is the vocation that is given 'the church'. Indeed, Word becoming flesh is the very process by which scripture comes to us in the first place.

The implications of this are huge. This becomes clear when in Paul's second letter to the Corinthians, he explains: 'in Christ, God was reconciling the world to himself, […] and entrusting the message of reconciliation to us' (2 Cor 5:19). Once again, he is saying the same thing as the Gospel of John when it quotes the resurrected Jesus as saying: 'As the Father has sent me, so I send you' (Jn 21:21). God's work of restoring all things to their right relationship, of healing all the cracks and divisions in the world, is now our work; and God's method of performing this work is our method. The medium truly is the message. *We* are called to be the message.

The body of Christ then, is that community of people who have given their lives willingly to be the hands and feet of God's work of healing a broken world; it is the vehicle of the Spirit of Christ in the world.

Paul has a profound understanding of God's purposes in choosing this way of working in the world; a way which is by nature plural and diverse. 'For as in one body we have many members, and not all the members have the same function, so we, who are many, are one body in Christ, and individually we are members one of another' (Rom 12:4-5). In his celebrated discourse of 1 Corinthians 12, Paul is at pains to emphasise that the 'weaker' and less well regarded members of this body are actually essential to its health, function and purpose. In other letters Paul insists that the body of Christ demands the obliteration of social divisions: 'There is no longer Jew or Greek, there is no longer slave or free, there is no longer male and female; for all of you are one in Christ Jesus.' (Gal 3:28)

In other words, the body of Christ is itself to embody the reconciliation, the healing and wholeness, or more accurately, the holiness, that is intended for humanity. As Stanley Hauerwas is fond of saying, the church does not have a social ethic, the church is a social ethic. It should come as no surprise then, that if the church is a social ethic, then its social ethic must also include its economic life. How can it not? So what does the economics of the body of Christ look like?

Perhaps the most important thing that needs to be said about the economics of the body of Christ, is that it is to be profoundly different from the economics of the world. In the New Testament, as in the Old, the call to holiness is a call to non-conformity:

> Therefore prepare your minds for action; discipline yourselves; set all your hope on the grace that Jesus Christ will bring you when he is revealed. Like obedient children, do not be conformed to the desires that you formerly had in ignorance. Instead, as he who called you is holy, be holy yourselves in all your conduct; for it is written, "You shall be holy, for I am holy." (1 Pet 1:13-16)

The last part of this passage is a quote from the Torah (Leviticus 19:2) – the blueprint for the Promised Land' – and it is the refrain that explains the detailed vision of Israel as an alternative economic community. Thus, the call for the body of Christ to embody an alternative economic ethic is not new to the New Testament, but is the calling that has been given to God's people all along.

The centrality of a new economics to the body of Christ is made fully evident in Acts chapter 2, the account of the birth of the church. The chapter begins with the dramatic arrival of the Holy Spirit. In the NIV Bible this is described as 'tongues of fire that separated and came to rest on each of them' (v.2). Ched Myers has pointed out the Greek in this text uses a very particular term – *diamerizo* – which is properly translated, 'tongues of fire were *distributed* among them'. This same term is echoed at the end of Acts 2: 'All who believed were together and had all things in common; they would sell their possessions and goods and *distribute* [*diamerizo*] the proceeds to all, as any had need.' Thus the story that narrates the beginning of the church starts with a distribution of the Spirit and ends with a distribution of goods.

What happened in the Jerusalem community in Acts was not communism – relinquishing of private property was not a condition of membership of this community. What happened was much more profound; as Acts chapter 4 explains 'no one *claimed* private ownership of any possessions'. What had changed was their attitude and outlook, or to paraphrase Paul, they had been transformed by a renewing of their minds and so no longer conformed to the patterns of the world (see Romans 12:2).

Paul makes clear in his letters that the transformation that comes with receiving the Spirit of Jesus is not some private transaction with God in our hearts, but one that must find its expression in a transformation of our day-to-day lives, meaning all of our acts of production, consumption and distribution. In Romans 12 he describes offering our 'bodies as living sacrifices' – that is, living each day, in the nitty gritty stuff of life, as a counter-cultural expression of God's love in the world. It is this day-to-day non-conformity for the sake of love, Paul explains, that actually constitutes our true worship of God (v.2).

It is in 2 Corinthians 8 that Paul most fully describes his understanding of the operation of the economy of God within the body of Christ. As within the Jerusalem Community, Paul sees this economy as being based on the freewill ('eager') and continual circulation of abundance towards need – 'The aim is equality' (v.14). Not surprisingly, Paul's understanding is explicitly grounded in the story of the manna in the wilderness where "The one who had much did not have too much, and the one who had little did not have too little" (v.15).

It is self-evident to the New Testament writers that participating in the community of Christ's spirit leads naturally to a transformation of our economic conduct. As the writer of the first letter of John plainly puts it: 'How does God's love abide in anyone who has the world's goods and sees a brother or sister in need and yet refuses help?' (3.17). As with Israel in the Old Testament, so with the body of Christ in the New Testament; any community called to witness to the character of the God of the Bible must be an alternative economic community. It is a community whose life is to be characterised by stubborn non-conformity to the ways of the world, and determined by a distinctive, lively and active vision of goodness, health and wholeness that embraces all of humanity and creation. In short, it is to be a holy people.

Which brings us back to where we started – the widespread and deep ambivalence about the church. For it will be clear to any reader that this vision of the body of Christ in the New Testament bears little resemblance to the church of today. There is not much that is distinctive about the economic lives and habits of Christians and Christian communities today. There is not much recognition that the sorts of choices we make in all our major economic decisions – choices about work, perceptions of needs and especially income needs, consumption patterns and investment choices – form an important part of the very substance of our following Christ in the world; that they are an opportunity to embody the ethic of the kingdom of God, an opportunity for the Word to become flesh. At best, there is perhaps a vague recognition that Christian ethics dictate that there may be some things we shouldn't do in these fields, however, the main field of decision-making is left open to the all-encompassing influence of modern consumer culture. And if we think about the challenges posed by climate change, or global food system, or about the dysfunctions of our cities – all products of an economic system – we find that Christians too are deeply implicated in the destructive economics of our time. Where did we go wrong?

Facing up to the crisis of the church

If we take seriously the great distance between the above two paragraphs – what we are called to be, and where we actually are – then it should not really be surprising that the church in the West is in crisis. Nevertheless, I do not believe that the church in Australia has yet fully faced up to this fact. Is it really in crisis?

There are a number of sources of information about the level of affiliation to Christian faith and the health of Christian communities: census data, National Church Life Surveys, and other academic studies.

These all look at slightly different things and employ different methods, however, they all confirm what anyone who is 60 or over could tell you – that identification with Christianity, and especially attendance at church, collapsed dramatically in the 1960s and has been in steady and continual decline ever since. In the last decade in Australia the extent of this decline has been masked by a numerical top-up through migration, however the continuing underlying trend is not in doubt.

If these trends continue as they have for the past four decades, then a number of mainline denominations in Australia will struggle to remain viable by the middle of this century (perhaps sooner) and they will certainly no longer be able to economically support the structures that we have come to associate with the institutional church. A senior leader in one of the mainline denominations once candidly shared with me that he had come to understand his job as one of 'crisis management', although he quickly acknowledged that few others in his denomination would be comfortable with such a description. Even in the religious culture of the US the decline of the church is biting hard; more than a decade ago Tom Sine wrote: 'If the present trends continue uninterrupted these denominations will be totally out of business by the year 2032'.

But what about Pentecostalism and the mega-churches? Aren't they growing? There is no doubt that the most significant phenomenon in the Australian church in the past two decades has been the growth of what are generally referred to as Pentecostal mega-churches (although the term 'Pentecostal' is becoming increasingly problematic when applied to many of these churches). Although these churches have individually grown very rapidly, they have brought little growth to the church as a whole. For the most part, their growth has come out of 'recycling' – picking up Christians who have fled smaller churches in the mainline denominations. But even here their growth is somewhat illusory – the mega-churches have very high flow-through rates (new attendees are largely matched by high rates of exit) and some observers close to this movement believe that the phenomenon has now passed its peak. Despite the enormous resources and effort poured into attractional-style services (more like concerts really), even these churches are struggling to win any *new* adherents to Christianity. As one sociologist of religion in Australia put it to me: 'Statistically speaking, conversion is a myth – it simply doesn't happen.'

However, such crude statistical trends conceal the fact that there are a couple of complex processes behind numerical decline, and we need to be careful to differentiate them. The first, and by the far the largest contributor to this numerical decline, is the end of the church's long historical cultural supremacy in the West. But although it poses a serious headache for the economic viability of many Christian institutions, we should not be too quick in lamenting this changing social position for the church. As shall be discussed later, the end of Christianity's honoured position within Western culture is perhaps a necessary precursor to a renewal of the dynamism of Christian faith.

It is when we consider the 'dynamism of Christian faith' that there is a serious warrant for concern; for it seems fairly evident that parallel to declining church numbers there has also been a decline *in confidence in the Christian gospel* amongst those who remain. This is a phenomenon which is plainly evident to most people in leadership in the church – indeed I would argue that it is perhaps equally present *within* the leadership – but it is much harder to demonstrate statistically.

However, there is one area where it does become statistically evident, quite alarmingly so, and that is in rates of the transmission of faith from parents to children. At this point we get a simple and yet profound insight of what the faith of parents has *meant* and *communicated* to their children, and it is a worrying picture. A study of religious affiliation in the UK, where the trend has been virtually the same, put it like this:

> … in Britain institutional religion now has a half-life of one generation, to borrow the terminology of radioactive decay. The generation now in middle age has produced children who are half as likely to attend church.[1]

---

1. David Voas and Alasdair Crockett, 'Religion in Britain: Neither Believing nor Belonging' in *Sociology* 2005 39:11 (Sage Publications), p.12

More precisely, the study found that children in families where both parents could be described as 'committed Christians' had only a 50% chance of sharing such a faith in their adult years, while children in families where only one parent was a 'committed Christian' roughly had only a 25% chance of sharing that faith. It is likely that this rate of decline is faster in Australia and if so, then this is undeniably a crisis.

In the midst of the pace and glare and noise of a hyper-modern consumer culture with all its demands, distractions and temptations, and all it moral and ethical upheaval, it is evident that the formulations of the gospel that have been the currency of the church for so many generations are now seeming increasingly *thin* and two-dimensional. Christian faith, as it has come to be understood, has become a mere accoutrement to life, easily dispensed with when life circumstances render it inconvenient, which they frequently do.

For some, this seeming deficiency in substantive relevance is compensated for by latching onto the passion and purpose of the social justice elements of the gospel that have so much to say in our current times. But, if you dig a little deeper it is not uncommon to find that this passion is still accompanied by a deep ambivalence about the rest of 'the package', and that social justice Christianity often looks little different from its secular counterparts. The growing recovery of the social justice purposes of the gospel (as they are ordinarily understood) is undoubtedly a positive development in the Australian church, but it is not proving to be a force for the renewal of faith. Those who buy this brand of Christianity are just as likely to drift out of the faith as not to.

Essentially, there is a crisis of confidence in the *goodness* of the good news. While many Christians are holding on (with varying levels of success) to a conviction that there is something deeply 'right' about the gospel, there is a great struggle to define, articulate, and therefore, to live by, a clear understanding of either what is distinctive about the biblical story or why it is good for us. And that is because the biblical story is not the only story we are listening to. There is a much louder story that has, for a long time, been filling our ears, and we have barely recognised that it exists.

The turnings of the church

To properly understand the current state of the church in the West we need to go back into its history to look at the things that have shaped it, and to ask where other stories began to be mixed up with the biblical story. To do this we need to re-visit the earliest centuries of the church, and there are perhaps many things that we could point to; but here I want to highlight four great turnings in the history of the church, which together go a long way to explaining our current predicament. I should quickly stress that these turnings all played out over generations and centuries and were never quite so clear as such at the time (lest we be arrogant about our superiority to our forbears). Also, in dealing with almost two millennia of history in such brevity, it is inevitable that this account will be too simplistic, and that every step of the way deserves more nuance and qualification. But it is critical that we try to gain some understanding of the grand narrative that has brought us to where we are. This is an inadequate first step, but as GK Chesterton once said, if a thing is worth doing, it is worth doing badly.

The first turning that I want to consider is the profound and incredibly unlikely revolution that gripped the ancient Mediterranean world when in 312 AD Constantine marched into Rome under the banner of the *Chi Rho* and began the process that would transform the Roman Empire into a Christian Empire by the end of that century. Depending on your theological view of history, this moment has tended to be seen as either Christianity's greatest triumph or its greatest calamity. I do not believe that such simplistic categories can accurately represent what happened, and why should they? The entire biblical story confirms that God is continually engaging in the messiness and mixed nature of human history.

We live in an ahistorical culture that is largely ignorant of its debts to its forbears. Much of what we most cherish most about Western civilization – our conceptions of equality, liberty, universal human rights and the inestimable value of every human being – are the hard won accomplishments of what is now disparagingly referred to as 'Christendom'.[2] Nevertheless, we needn't diminish these priceless gifts to humanity when we also acknowledge that when whole Roman Empire became 'Christian', something else profound happened to Christian faith. As John Howard Yoder points out, before Constantine it took deep courage and conviction to be a Christian; after Constantine, it took deep courage and conviction not to be a Christian. When the church took up the sword there was little room left for the faith of the cross and the Sermon on the Mount. This was the first great turning. From this time on and until very recently, the whole history of Europe and Western civilisation has proceeded on the assumption that it was a Christian civilisation.

The next great event I want to consider here is one of the unforeseen consequences of the Protestant Reformation, which began with Martin Luther and John Calvin in the 16th century. In many ways the Reformation was a counter-reaction to the Constantinian Legacy that made everyone by default a Christian, and which made the institutional church the monopolistic arbiter and mediator of faith for everyone. The Reformers demanded that faith be a matter of conviction and integrity, and insisted that every individual human being could have access to God without the mediation of the church. This was the brilliant light of truth at the heart of the Reformation, however, as RH Tawney said, it was a light almost too blinding for the Reformers. As future generations of Calvinists and Puritans worked through these ideas the ultimate result (which neither Luther or Calvin would have been happy about) was an intense focus on salvation of the soul after death, and the obliteration of the older, deeper understanding of humanity as a creature of the earth, made for communion within the fellowship of creation and the Creator. Reflecting on later generations of English Puritanism, Tawney concludes that 'it enthroned religion in the privacy of the individual

---

2. For a lively account of the accomplishments of 'Christian civilization', see David Bentley Hart, *Atheist Delusions: The Christian Revolution and its Fashionable Enemies*, Yale University Press, 2009.

soul, not without some sighs of sober satisfaction at its abdication from society'.[3] This was the second turning.

At precisely the same time as the Protestant Reformation (and feeding directly into it), European civilisation was undergoing another monumental transformation, which is the subject of our third great turning. Between 1450 and 1650 a new economic system was coming into being, which we today refer to as capitalism (although this term was not coined for another 400 years). This is a word that gets bandied around a bit, and depending on who is using it, carries either negative or positive connotations; however, there is very little clarity about what capitalism actually is and how it differs from other forms of economic organisation. (For example, many people assume that capitalism is synonymous with 'free enterprise', which it is not; you can have capitalism without free enterprise and free enterprise without capitalism.) Let it suffice for the moment to stress that the form of economic organisation being birthed in Europe was something entirely new and something that radically reshaped the social landscape wherever it took root. The end result was that all of society, from milk-maid to monarch, was made to serve a new Law and a new Lord: the Law was that of never-ending accumulation and constant growth; and the Lord was Capital, better known as Mammon.

Capitalism went on to conquer the globe and become the only economic system going, and all the while there remained the assumption that it was the product of a Christian culture. The post-Reformation church, which in centuries-past had doggedly applied the gospel to the economic life, offered little resistance to this economic upheaval, and very soon the church had baptised the virtues of efficiency, productivity and accumulation as the new 'godly discipline'. This was the third turning.

Hard on the heels of the Reformation and the birth of capitalism there came that other profound revolution in European society, this time intellectual, known as the Enlightenment. It is far beyond me to try to summarise the diverse works of philosophy and science emanating from thinkers such as Descartes, Locke, Newton, Voltaire, Rousseau, Hume

3. RH Tawney, *Religion and the Rise of Capitalism*, Penguin, 1938, p.228.

and Smith, but there is no doubt that it set in motion a profound shift in human thinking that is still being worked through to this day. Once again, we should resist the temptation to arrive at simplistic positive or negative assessments of such a complex and profound historical movement, but we do need to look more critically at what is otherwise widely perceived to be an unqualified leap forward for humanity.

Immanuel Kant summed up the spirit of his age as simply the freedom to use one's own intelligence, but it was more than that. Emboldened by the Reformation's relegation of the church to a secondary sphere in society and the banishment of God to a spiritual sphere (if he existed at all), the Enlightenment project set about trying to understand the universe and human existence without any recourse to what was widely becoming viewed as 'superstition' – that is, religion or the church.

The Enlightenment set about to establish the self-sufficiency of human kind and that is what it accomplished. The church was graciously granted its domain to chaplain the private recesses of the human heart, but it was made clear that it had no place in the realms of politics or the new science of economics – these were simply no longer of religious concern. The church basically accepted and agreed with this position – the fourth great turning. Through all of these storms and upheavals, the church, the government and even the new breed of non-Christians, continued to assume that they lived in an essentially Christian society, even though everyone had been transformed into practical atheists.

To sum up then, we have considered four great turnings away from the gospel by the church:

1. We consented to join with empire and domination, and 'Christian faith' became deeply confused with the culture of European civilization.

2. We individualised and spiritualised the Christian gospel, effectively discarding all that it had to say about our material lives and the bonds of community.

3. We became willing participants in the program of never-ending accumulation.

4. Other than a narrowly defined spiritual sphere, we made God practically irrelevant in virtually every other sphere of life.

Of course, this is not the whole story. There are indeed other turnings in Church history that would be fruitful to consider if space permitted. But, perhaps more importantly, there has always been a counter-story, a story of faithfulness, resistance and hope at the margins – the desert fathers, the early monastics, the mendicant friars, the Anabaptists, the Diggers and Quakers, the list goes on – and their legacy continues to be renewed and live on.

But, as the inheritor of the four turnings discussed above, it is not really that surprising that the church now seems disoriented and irrelevant, like a house built on a foundation of sand. In the face of global environmental crises, unprecedented inequality, economic tumult, social fragmentation and the hollowing-out of self, we are discovering that it all looks very little like the good news that Jesus proclaimed, and that it has all been a profoundly destructive project.

The economics of renewal

To put it bluntly: the crisis of the church is one and the same as the crisis of the world, because that is what we have become. Whether it be the global financial crisis, climate crisis, resource depletion crisis, mental health crisis, or the crisis of the family, Christians find themselves implicated like everyone else. We have lost sight of the gospel's call to a radically distinctive way of life that rejects the programs of power and wealth, and refuses to participate in activity that does harm to another or does harm to the earth, which is the same thing.

If the church is in crisis in a time of general crisis, what would renewal of the church look like? What would it take for the church to begin to more fully reclaim its original vocation? What would the church look like if it were to be a distinctive community that bears *good news* for a world in crisis?

It has been oft observed that times of crisis are also times of opportunity. The prophet Jeremiah, during a time of deep crisis for his people, at one point interrupts his dark warnings to offer this cry of hope:

> Stand at the crossroads, and look, and ask for the ancient paths, where the good way lies; and walk in it, and find rest for your souls. (Jeremiah 6:16)

Jeremiah understands that if hope exists anywhere, it does not exist in denial; rather, the beginning of hope is being brought to a standstill. It is only from this place that we can look clearly and soberly at the truth of our predicament, and it is only when we come to such a place that we can see clearly that there is another possibility - we can choose a different road. It is instructive that for Jeremiah, finding the way of hope and goodness does not require making new discoveries, but upon *remembering* what has been forgotten.

Earlier I cited some of the statistics that indicate the church's social position is in radical decline. This tells us that, whether we like it or not, the shape and structure of the church in the coming years will no longer be what we have known it be. However, this should not be interpreted simply as a negative thing. It is almost certainly necessary that Christianity fall from its place of cultural privilege in the Western world in order for it to begin to disentangle itself from those great narratives that lie at the heart of Western civilisation – individualism, capitalism, secularism – and that have so confused our understanding of the gospel of Jesus. The more penetrating Christian thinkers have seen this truth for a long time now. Visser 'T Hooft, a European Christian leader, wrote this in the 1930s:

> What is the Christian task at this crucial moment? In the very first place it is necessary that Christians understand the new position, which is completely different from the old one. They must realise that they can no longer count on the momentum of the old tradition, that they are no longer going to be treated as the honoured representatives of the main current of culture, or, to put it quite shortly, that they will be less and less at home in the West. The West is again becoming for them what the Roman world was for the early Christians: *a world*

whose presuppositions contradict their faith, a world which is not only secretly but quite openly indifferent or even hostile to their essential convictions.[4]

The more clearly we begin to see the divergences between the gospel and mainstream culture, the more conscious we will become of our need for renewal. It would be irresponsible of me to suggest that there is some formula for renewal, and delusional for me to try and make a prescription for it. That said, perhaps there are some things that we can point to that may be necessary if the church is to be rejuvenated as an alternative community that witnesses to a new human possibility.

Surely, any movement of authentic Christian renewal has to be centred on a rediscovery of Jesus and his message, and the expression of that discovery in people's lives. A central element of this – but by no means sufficient – must be the reclaiming of the biblical story's distinctive perspective on our material lives. In a time when the bad news confronting humanity centres on the structure and content of our material lives, the good news of Jesus will only fully become good news when it also finds expression in our material lives.

What might such a movement look like? Let me suggest seven dimensions of an alternative economic life that would express God's counter-cultural good news in 21st century Australia. Here I will focus primarily on the individual lives of Christians, rather than address the economics of the institutional church – in this matter, there can be no real or lasting transformation of the institution until change is embraced and embodied by its members.

1. Rejection of the idols of 'more' and 'me'.

Perhaps the greatest task to be undertaken is to become conscious of, name, and then resist, the operation of two great false gods in our time – consumerism and individualism. These two forces go far beyond the cultural habits and practices of shopping – they shape our very deepest convictions about what life is for, and what a good life is. Our

---

4. Visser 'T Hooft, *None Other Gods*, SCM Press, 1937, p.106 [my emphasis].

understanding of everything – even family and work – is refracted through these lenses.

The first part of this struggle is mental and spiritual. It requires becoming conscious of the ways in which we are shaped – more than we know – by the ocean of advertising and marketing that we swim in, and by the example of those we see around us. We need to take account of what is shaping our minds and our hearts, and take some responsibility for what voices we give our ears to.

The second part of this struggle comes down to some big, life-shaping, practical decisions. What sort of income do I and my family need for a decent life? What sort of home and material possessions do we require to be satisfied? How much is enough? However we might answer these questions, it seems fairly safe to say that the poor of the world, our families and communities, and even the earth itself, need us to answer: 'We can live with less!'

If Christians could, *en masse*, make active, conscious and careful choices to live with less, it might open up all sorts of new possibilities. Certainly, it is hard to think of a more counter-cultural or evangelical witness than this. And not only might new things become possible for Christian communities in all sorts of practical matters, it might allow spiritual breakthroughs which cannot take place while these two idols hold sway.

2. Care & nurture

In place of conforming our lives to the narratives of 'more' and 'me', we need to pass our whole lives – family and relationships, work and leisure, consumption – through a new filter, and ask a new fundamental question: how does this further God's great work of restoring a broken humanity and a broken creation (see 2 Cor 5:18-20)? How does it bring forth goodness in the world?

Another way of stating this is by testing all of the components of our lives against the simple ethos of care and nurture. What is the effect of my choices (whether it be work, consumption or recreation) upon others

– upon my family, upon my community, upon the poor and vulnerable, upon the earth and its creatures, and upon my connection to God? How can I take care for them in my choices? How do I actively nurture my family, community, and the earth itself?

3. Work

Rejecting the idols of 'more' and 'me', and shaping our lives to the ethos of care and nurture, demand that we radically re-think our attitudes to that activity which dominates most of our waking lives – work. This is a subject which deserves much fuller elaboration but here let me articulate some critical questions:

- Why do we work?
- What are we seeking out of work?
- How much income do we need from work?
- What work needs to be done?

To think well about these questions we need to include in our frame of reference all of the unnoticed and unpaid work – most of it work of care and nurture, much of it still shouldered by women – that upholds families, communities and society itself.

Rather than simply thinking about work from the perspective of income and individual fulfilment, we would do better to think first from the perspective of the household: what work – paid and unpaid – needs to be done; how much income is needed; who does what, and how do all parties find dignity and worth in their work? Rather than work being an activity that isolates members of a household from each other, can the working patterns of the household become ones that reflect mutual care and nurture of each of its members, adult and children, and which are owned and valued by all? When it comes to choosing paid employment, how can we choose employment that serves the world rather than exploits it?

These questions are particularly pressing for those younger ones who have not yet solidified major life decisions about work, especially for those who have the real luxury of choice (not everyone does). William Temple, Archbishop of Canterbury during the Second World War, took a very serious view of the responsibility such choice confers:

> Some young [and we could add, older] people have the opportunity to choose the kind of work by which they will earn their living. To make that choice on selfish grounds is probably the greatest single sin that any young [or older] person can commit, for it is the deliberate withdrawal from allegiance to God of the greatest part of time and strength.[5]

For those already well down the path of work decisions, making changes can be much harder, and we need to be realistic about what is possible in the short term. Even for these, though, in the medium to longer term, there may well be more possibilities for change than is generally acknowledged. What is certain is that the more we can make a choice to live with less, the greater degree of freedom we will have to make all sorts of creative choices that nurture goodness in ourselves, our families, communities and the land.

4. Responsible consumption

In modern Australia, most of us have been relegated to the position of consumers – very few of us actually produce anything tangible and material, let alone produce things that serve our basic everyday needs. Moreover, we sit at the apex of a global consumer economy that has been so structured to ensure that we can devour new consumer items – that is, the earth's resources – at an ever increasing pace. It is a profoundly destructive system, and if we are going to be people who work for healing and wholeness we must begin to take responsibility for the impact of our consumption.

At the heart of taking responsibility for our consumption is the preparedness to *pay more* for many of the things we buy. We can afford to consume so much because we do not pay the true cost of things – either to the earth, or those involved in the making and getting of things

5. William Temple, *Christianity and the Social Order*, Penguin, 1941.

to us. By paying more – for fair trade, organic, locally sourced, more ethical, less wasteful, less disposable, more durable, more recyclable, less plastic, less toxic etc. – we are expressing a higher level of care for our neighbours and for creation. Of course, paying more means we would not be able to buy so much – we would need first to accept living with less ...

Finally, we cannot talk meaningfully about responsible consumption without addressing technology consumption. The cumulative effect of global mining, ever-cheaper labour, and e-waste that underpins this industry is vast, and that is not saying anything about the damage we are doing to ourselves, our families, our communities and culture with our mindless devotion to novelty and distraction. Somehow we need to get off the ever accelerating treadmill of constantly purchasing and updating new gadgets, and treat technology purchases with a seriousness that will be incomprehensible to the broader culture.

5. Household economy

The corollary of rediscovering a healthier approach to work and becoming more responsible in our consumption is rediscovering the household as a place of good work, even a place of production, for both men and women. With the advent of the industrial revolution, men became alienated from the household economy filling only the role of bread-winner, and women bore the burden of care alone while increasingly finding this role was ascribed no social value. Now the general consensus is that everyone should aspire to escape the 'drudgery' of home-making; we have been taught to think of our homes merely as places of comfort, retreat and ever increasing luxury, and all of us are the lesser for it. Wendell Berry writes:

> The modern failure of marriage is a failure of our home economies. The practical bond of husband and wife in the home has almost disappeared. It's a sacred and practical bond that gives order to a home, to family, to their descendants and to community. The work of the home is the health of love. And to last, love must enflesh itself

in the material world – produce food, shelter, warmth, surround itself with careful acts and well-made things. [6]

If we are to take more care in our consumption and if we are to live with less extravagant budgets, this will take serious thought, time and creativity to be applied to the household economy, and it will require both men and women to apply their skills to it, and to re-learn many skills that have been lost. The work and challenge of nurturing healthy (that is, holy) households is a life-giving adventure waiting to be rediscovered.

6. Generosity

Not only does the Biblical message insist that God's people forsake activities and ways of life that gratify the self at the cost of others, it consistently calls us to give money away! The call to generosity, to share out of our abundance, is not simply a call to charity and philanthropy, it is a central and practical expression of the deeper spiritual movement that the gospel calls us to. At the precise time when we are wealthier than ever before, our levels of generosity are declining. We have lost the habits of structuring generosity at the centre of our household economy (tithing) and we have become hardened to that impulse which responds spontaneously to need without calculation (almsgiving). Relearning these practices is not just a matter of social responsibility; it is a requirement of spiritual health.

At this point it is worth pausing to do some quick sums. So far, I have said that we need to learn to live with less, to be prepared to pay more for what we buy, and to be more generous in giving money away … what!

Live with less + pay more + give generously = ???!!! *(Are you serious?)*

Let's be honest: as far our culture is concerned, this is a ludicrous suggestion. It is a scandal to the Jews and foolishness to the gentiles, which is a clue that there might just be something of the wisdom of God in it.

---

6. Wendell Berry, 'The Body and the Earth' in Wendell Berry, *The Art of the Commonplace: The Agrarian Essays*, Counterpoint, 2002, p.118.

7. Economic interdependence

Up to this point I have discussed ideas concerning individuals and households. The current of our time is towards the disintegration of community, making our relationships with others increasingly abstract - dependent no longer upon locality or necessity, but upon nothing more than choice, making friendship yet another variant of consumerism. The opposing current then is to begin to re-establish some practical and material interconnections with others, and especially within Christian communities.

I am not here suggesting that we try and emulate the Jerusalem Community of Acts chapters 2 and 4 – that is a story which is a sign in the distance, calling us forward. However, there are many small and modest initiatives of sharing and cooperation – such as sharing mowers, whipper-snippers and trailers, making soap together, neighbourhood gleaning and salvaging, or running food and bulk-buying cooperatives – which really can practically enhance our weekly lives, and begin to add a new dimension to our connections with others. Once we become better at such small expressions of economic community, who knows what bigger things might begin to seem possible?

I should conclude with some qualifications about what I have said above. Once again, I want to stress that these ideas could never be considered as a formula or prescription for renewal of the church. I do believe that some sort of movement to reclaim the radical and counter-cultural implications of the gospel for our material lives will be a necessary element of any recovery of deeper and more authentic expressions of Christian faith and community. However, other movements, deeper spiritual movements, will be needed too, and who is to say what will be cause and what will be effect?

Secondly, such a brief description of such big life movements is clearly unable to do justice to the complexity and practical difficulties involved in any of the movements described above, or to engage the huge variations in people's living circumstances that form the starting points from which any of us come at these issues. Moreover, most of these movements cannot be implemented over-night, or even in a year – they require the work of small but consistent steps over many years, perhaps the rest of our lives.

What I have articulated here is the merest beginnings of a sketch – it needs so much more colour and detail. And let it never be said that I have suggested any of these movements is easy, let alone all of them together. But I will contend that they are all perhaps more possible and more attainable than our cultural programming would have us believe.

What I am proposing here is not a heroic lurch to some radical new experiment in community – something likely to be a flash in the pan. Rather, what we need to recover is *a biblical lens on life* – eyes to see and ears to hear – and to begin, from the contexts where we find ourselves, to re-examine the thousands of choices we make through this single lens. The movements that will be of most significance will be ones made carefully and thoughtfully, with a clear view of the complexity of life. Whether we are making big steps or little steps, what is important is not so much what we succeed in attaining, but the direction in which we are travelling. The trajectory of our lives needs to be clear enough that our children will comprehend it and take the journey further than we are able.

Jeremiah calls for us to make a turning at the crossroads and follow an ancient way. But if this road seems difficult, obscure and little trodden, then let us take comfort that this is the way of Jesus:

> For the gate is narrow and the road is hard that leads to life, and there are few who find it. (Matt 7:14)

It is a road that can only be walked one step at a time.

SHAKING THE HOUSE
Being the Household of God in the midst of dangerous climate change

"Yet once more I will shake not only the earth but also the heaven."
Hebrews 12:26

'For the time has come for judgement to begin with the household of God'
1 Peter 4:17

'The rain fell, the floods came, and the winds blew and beat on that house,
but it did not fall, because it had been founded on rock.'
Matthew 7:25

As I begin writing this, it is a bright and sunny day in the early winter of Bendigo and my spirit cannot help but be lifted by the scene before me. The 'Autumn Break', as the farmers call it, has brought good rain this year, and the rust and yellow hues of late Autumn are set-off by a green lushness which heals the barrenness of a hot, dry summer. My daughters have gone off to school with a similarly bright and sunny attitude to the day ahead, and, implicitly, to their lives ahead. As my mind turns to the future, my deepest desire is to say to them the words received by Julian of Norwich, "All shall be well, and all shall be well, and all manner of things shall be well". Somehow, I believe there is a deep truth in this, and yet … and yet I also feel a deep disquiet about what lays ahead for my children. All manner of things are not well. I am painfully conscious that, somehow, I must help them to look clearly into the future that awaits them, and help them find a place to stand, a vision to live by, and a hope to hold onto in a darkening world. Because that is almost certainly what confronts us.

At the heart of this essay on climate change is the very difficult task of trying to look into the future to ask what it means for us now. And certainly the message that is coming with depressing constancy from the scientific community is that all manner of things are not well, and indeed the scenarios being discussed are truly frightening. When this is married with a clear-sighted diagnosis of political will (in this country and abroad), the power of vested interests, and the deep psychological resistance of the general public, then it is very hard to be optimistic about our capacity to avert seriously dangerous climate change – the sort that will unravel the world as we know it. There is a great chasm between the economic systems which we humans have come to see as our birthright and the needs of the biosphere. The movement to reconcile this dissonance between the two has been, to use an unfortunate metaphor, glacial.

Of course, the plain fact is that we simply do not know what the future holds. They say that hope springs eternal, and perhaps there is more cause to have hope in the unknown than there is in what we think we know. But the condition of humanity has always been that we must make decisions about an unknown future based on what we know now. And the message of the Bible has always been that we know *enough* now to choose for life – not just our own little lives, but for the whole Community of Life in which we share. If only we can remember.

It is my concern here to ask more pointedly what the challenge of climate change means for those of us who would be followers of Jesus. Just as the changing climate of this century will likely be a crunch-point for humanity, it is in many ways a crunch-point for Christianity too, especially in the West. The exasperation of climate scientists is very closely paralleled by those in the church who are looking clearly at the signs of the times and can see that the writing is on the wall for the future of mainline Christianity this century. The powerful inclination towards business as usual, to enter into an everyday practical denialism, is as evident in both the membership and leadership of the church, just as it is evident in the broader community with respect to climate change. Indeed, the climate crisis and the crisis of the church are not merely parallel phenomena, they are both manifestations of the same greater crisis – the crisis of the way that we live.

My core contention is firstly that the good news of Jesus does indeed offer a light and hope by which to navigate the dark times that seem to lie ahead, and secondly that this hope is foundationally connected to how we live, and not just what we believe. I am suggesting that it is, and always has been, fundamental to the vocation of God's people to embody and proclaim the reconciling of economics and ecology which is at the heart God's great work of restoring all things, and at the heart of this work is the intimate and practical reconciling of economics and ecology within our own homes. Indeed, the Christian community should be particularly well equipped to speak with wisdom and insight into the challenges now posed by climate change - it has always been the core business of Christianity to bring the future into the present, particularly with a view to instructing how we ought to live now. We should be, but we are not. Before the Church can begin to more fully reclaim its vocation, it must go through a profound shaking of its own, 'For the time has come for judgement to begin with the household of God' (1 Peter 4:17).

The climate crisis and the crisis of the church

Perhaps the only thing that is certain about the future is that we cannot be certain about much. As Wendell Berry has long urged, it is about time we became a bit more honest about our ignorance. The more we learn about the Earth's climate systems, the more we glimpse how mind-bogglingly complex they are, and how little we know. Nevertheless, there is now a vast weight of evidence that tells us that climate change is proceeding at a truly alarming rate, and while we can make guesses, it is almost impossible to predict the complex spin-offs and feedbacks in our ecological and climate systems this may set in motion. As far as I can tell, there seems to be a sombre consensus among climate scientists that we have probably (with due deference to what we do not know) missed the window of opportunity to avert dangerous climate change; the real question now is how bad it will be.

Many of those advocating a radical de-carbonisation of the economy have suggested that our actions now, or failure to act, could lead to a number of possible scenarios for the unfolding century. There are variations in how these scenarios are presented, but they can basically be summarised thus:

1. We could achieve a relatively smooth transition to a low-carbon economy, keeping climate change within a manageable range, and adapting whatever changes do take place, with some pain, but not too much trauma.
2. We could undertake a rocky descent into radically changing ecology and global economy, punctuated by widespread and repeated crises in the human population.
3. Human civilisation as we know it could collapse, along with much life on earth as we know it.

It now seems that the first of these possible futures has probably passed us by.

Let me stress that I do not raise this to suggest that we should now abandon attempts at climate change mitigation and move only to think about adaptation. On the contrary, it seems all the more urgent now that we pull out all the stops to cease humanity's extraordinary vomiting of greenhouse gases into the atmosphere so that we can avert whatever we possibly can still avert. Nor do I want to appear fatalistic or deterministic about our climate future – we truly do not know what the future holds. However, it also seems to me that any facile optimism about the capacity of humanity to 'find a way' will not be a help to us, and is merely another form of denialism. Given the enormity of the scenarios now being discussed, the coming century is going to require of us mental and spiritual resources which we do not yet collectively possess.

Which brings me to my other concern – the church. To those who care to pay attention, it has long been evident that the church in the West is in crisis. Not just the church, but Christian faith itself. My concern here is not so much with the much commented upon continual decline in church attendance and affiliation since the 1960s. This is to be expected as the last vestiges of Christianity's post-Constantine cultural hegemony works its way out of the system, and while it holds huge implications for the future economic base of denominational church structures, it also opens up new possibilities for grass roots renewal.

I am much more concerned about two clearly observable, but less often discussed, phenomena. The first is the breakdown of transmission of faith between parents and children: a study in the UK has shown that children in households with two 'highly committed' Christian parents only have a 50% chance of continuing the faith of their parents, and the rates are lower where commitment is not so clear.[1] The second is the crisis of confidence that is so evident amongst the existing membership of the church about the 'goodness' of the good news. The ethical, theological and sociological challenges of late modernity are so ever-present that many in the leadership and laity of the church aren't sure what to think or believe any more, whether on matters of theology or ethics.

What the Western Church is now experiencing is the inheritance of a gospel that for 200-300 years has been stripped of most of its material substance and guidance for living in the world of Caesar and Mammon. Instead of a faith whose central testimony is that the Word becomes flesh, modern Christians have mostly received a privatised and spiritualised faith which barely touches their day to day economic conduct.[2] The ethical choices of career, income, investment, consumption and leisure are barely recognised as such, but are rather merely arenas of 'normal life' where there seems to be little reason not to do as everyone else does. In the unrelenting and all-encompassing glare of post-Enlightenment secular consumer society, with its radical individualism and moral relativism, modern Christianity has seemed a thin and feeble thing.

Both the climate crisis and the crisis of the church are thus two facets (there are many more) of the one great crisis – the crisis of the way that we live. What we are now finding, and able to empirically verify, is that idolatrous economic systems destroy the natural order of creation, they wreck human community, and they impoverish the human spirit. The Bible has ever said it was thus, but more on that later. The key point

---

1. David Voas and Alasdair Crockett, 'Religion in Britain: Neither Believing nor Belonging' in *Sociology* 2005 39:11 (Sage Publications), p.12. A sociologist of religion in Australia has suggested to me that the rate of generational faith loss was perhaps a little faster in Australia.
2. See chapter 1 for an account of church decline and the long sequence of historical turns that brought about this state of affairs.

for the moment is simply this: when we are asking what Christian faith means for the challenges of climate change, we are not bringing together *two* foreign things, ill at ease in each other's company; rather, we are in fact re-joining *three* foundationally connected, though long separated, bases of human life: ecology, economics and faith. No one of these three things can properly be understood without understanding all three together. It turns out that many of the everyday alternative economic choices which we need to start making for the sake of the planet, are also choices which are likely to strengthen and deepen the life of faith, and visa-versa. But before we can positively begin to unpack this story, there are some deeper and darker obstacles that must be confronted.

Grief and fear

It seems to me that it is impossible to think well and clearly about the challenges posed by climate change – whether moral, theoretical or practical – until we have addressed the monumental obstacles of our grief and fear. It is now clearly evident that there is a deep psychological resistance within the community that is evident in the failure to really grasp the implications of climate change. This can take a number of forms, from the reflex instinct to deny it outright, through to a more complex acceptance of the science, but nevertheless somehow quarantining that knowledge so that it does not impinge on everyday life. I do not think any of these reactions can be properly understood without understanding the role of grief and fear.

I have noticed a number of phases in my own engagement with climate change issues. Initially, as is my wont, I leapt into campaigning for drastic carbon reduction with an activist ardour that was powerfully fuelled by the urgency of the challenge and the hopefulness that the window of opportunity was still open to us. Then a number of things happened: Australia's long drought ended and the general concern about climate change seemed to dissolve with the rain; the Copenhagen Summit failed; the fossil-fuel-funded 'merchants of doubt', spearheaded by the Murdoch Press and the Coalition, gained increasing traction in the public discourse; successive Labor Governments spectacularly combined moral cowardice with political idiocy; and then the Abbott Government was elected with a mandate for swinging a wrecking ball through the little progress that had been made. At the same time as all of this, the science coming from Greenland and Antarctica has been suggesting, with increasing conviction and urgency, that things may well be worse than we thought.

At some point it became clear to me that we have most likely missed the opportunity to avoid dangerous climate change, and the only question now is how bad it will be. My reaction to this, without ever consciously admitting it, was to effectively turn away from it all. It was all too much. The issues were too big, the implications unthinkable, and there was nothing I could do, so I may as well just get on with life today. This lasted a while, but as someone who is naturally in touch with these issues, I could only hold out for so long. I was eventually confronted with my avoidance, and that is when the grief and the fear set in.

Grief. A small word but a deep thing. It is not always easy to know exactly what is being grieved. Certainly, I feel grief for the diminishment of the natural world which I deeply love. I am a North Queenslander who loves snorkelling, so the prognosis for the Great Barrier Reef holds sharp pangs for me personally, as well as grief for the loss that this represents for my daughters and future grandchildren. According to the *Weekend Australian*, the University of Queensland's Global Change Institute has told a senate inquiry that under current conditions of bleaching and other stressors, 'the reef as we know it would be destroyed within a human generation'.[3]

These are powerful and palpable griefs, yet they are comprehensible. But there is more to it than that. Even more deep seated and harder to face is the complex grief of guilt, the realisation that *we have done this*. For me, the easiest resolution is to transmute this grief into anger and blame the failure of others, especially our forbears and our political and business leaders. But I know that is only half the truth. I, like so many

3. J. Walker, 'Coral comes back from the dead', *The Weekend Australian* (*Inquirer*,p.17), June 21-22 2014.

others, have participated in the binge of global travel that accounts for a surprisingly large part of our climate impact.[4]

And there is an even more deep seated and confusing grief yet. It is the peculiar and parochial grief of the Western middle-class. Whether Christian, agnostic or atheist, we all harbour a deeply held assumption *that bad things don't happen to us*. With the exception of car crashes and cancer, tragedy is the preserve of browner people in other continents, witnessed on TV screens. And yet we are now confronted with the fact that the way of living which we have come to assume as the teleological destination of humanity, can no longer continue. It turns out that, religious or not, we have all invested a deep faith in the idea of Progress, and, like all faith in idols, we have now been left with a bitter and empty hollowness. And the collapse of this elaborate mental and spiritual façade is absolutely terrifying. Grief becomes fear.

Fear. It has taken a while to admit to myself that I am now deeply fearful about the world that awaits my daughters as they grow up and begin to have families. There have been periods when this fear has had a visceral grip on my body, with sleepless nights, a tightening of the stomach and the heavy hand of depression. Certainly, I am deeply worried about the physical changes that climate change will bring about and the challenges for day to day living that this will bring about. We have recently moved to Bendigo, in an area that was affected by the 2009 Black Saturday bushfires. Having being raised in a culture that implicitly assumed that it was basically independent of the natural world, I now find that the biggest practical challenges facing us in the coming decades will be those of fire and water.

But, as a student of history and politics, I am even more worried about the chain of events that these sorts of changes around the globe will ignite in politics, society and economics, both within and between

---

4. Aviation accounts for 2% to 6% of global greenhouse gas emissions, however burning fuel at high altitude has nearly three times the climate impact of burning the same fuel at ground level. See Thea Ormerod & Miriam Pepper, 'What is it about plane travel?', *Manna Matters*, November 2013 <www.mannagum.org.au/manna_matters>

nations, and what it will all mean for a frighteningly interdependent and fragile global economy. I wish I did not know of the depths to which humans can sink, but I do know, and I cannot ignore it.

Some may feel that this is overly negative and pessimistic terrain that I am opening up. Among activists there can be a sense that we must stay positive so that we can keep campaigning, so that we don't put people off. But does there comes a time when facile optimism, or even facile hopefulness, only serves to dig the hole deeper?

And for Western Christians, this terrain all seems so foreign, so distasteful. Surely Christianity is a message of hope? How strange that we have not noticed that the foreboding of doom occupies a very large part of the terrain of the Bible! How can we understand the prophets if we do not understand that they have foreseen disaster approaching, the natural consequences of human sin, and yet they must watch their people plunge merrily and optimistically into the jaws of crisis, in denial of the inevitable until it comes to pass.

My brother once commented that Jeremiah is the prophet for our times, and now I understand what he means. There is a heart-rending moment where the prophet proclaims a powerful hope that disaster can yet be averted:

> Stand at the crossroads, and look, and ask for the ancient paths, where the good way lies; and walk in it, and find rest for your souls. (Jer 6:16)

But in the very next statement this hope is dashed: "But they said: 'We will not give heed'". Jeremiah is tormented by a grief that has multiple dimensions: (i) grief over the future that he has glimpsed for the people he loves; (ii) grief that he must be the one to speak of it to people who do not want to hear; (iii) grief that he has seen how doom can be averted, and yet has to watch the wilful denial of his people, the refusal to alter the course they have set for themselves: 'Your wrongdoing has upset nature's order, and your sins have kept away her bounty' (Jer 5:25).

It is clear that the Bible has fully and repeatedly anticipated the human predicament that we are now witnessing in relation to climate change. In these ancient pages there is already a deep understanding of the anthropology that underpins the great social psychology phenomenon of our time – our failure, despite overwhelming evidence, to heed warnings of impending disaster. It is only when we understand that the Biblical witness has fully looked into the face of human darkness, that we can properly understand it as a message of hope.

Suffering and hope

It might seem odd to say, but before we can recover a fully Biblical witness of hope, we must first recover its witness of suffering. It is an indication of the extent to which Western Christianity has become conflated with modernity and progress that many Western Christians assume that faith in Christ should mean things go well for them in life. Indeed, the experience of suffering can very frequently prompt a crisis of faith, so we can see what a colossal challenge to this type of faith climate change presents.

In the New Testament we find the opposite assumption about the link between faith and suffering:

> Beloved, do not be surprised at the fiery ordeal that is taking place among you to test you, as though something strange were happening to you. But rejoice insofar as you are sharing Christ's sufferings, so that you may also be glad and shout for joy when his glory is revealed. (1 Peter 4:12-13)

The specific context of this passage is the experience of persecution that the proclamation of Easter inevitably provoked from the Imperial order. However, it also reflects the much deeper recognition that irrespective of our faithfulness, irrespective of our hope in Christ, the world may yet darken; bad things may yet happen to us.

Indeed, we need to go much further than that. Again and again through the New Testament, in so many ways, the core identity of Christ's followers in the world is actually staked to that very experience of suffering in

46

a fallen world – 'you are sharing Christ's suffering'. When Paul said 'you are the body of Christ', he was articulating a deep identification of the community of Jesus' followers as the continuing incarnation of Christ in the world – those who, in the fullest sense, are sharing in and continuing Christ's purposes of reconciling a broken cosmos (2 Cor 5:16-21). And for Paul, the understanding of 'Christ crucified' was thus central to the church's self-understanding of its work and purposes in the world: 'I am now rejoicing in my sufferings for your sake, and in my flesh I am completing what is lacking in Christ's afflictions for the sake of his body, that is, the church.' (Col 1:24).

Jim Punton, a radical Church of Scotland minister and youth worker of the 1970s and 80s, used to teach that 'The gifts of the spirit are given to the people of God to give them strength to suffer'. But this has become a hopelessly alien message in our churches – there are few church leaders who would today feel comfortable emphasising this core thread of the Christian message. Nevertheless, once we begin to recover the Christian vocation of bearing the pain of the world, and if we take seriously Paul's language about the pain of creation (Rom 8:18-25), then we are forced to look at climate change in a whole new light; one in which the tragedy of climate change becomes even sharper, almost unbearably so, and yet this is precisely the unbearable cost of love for the world that we are called into. It is not for nothing that the second beatitude, following the promise of inheriting the kingdom of heaven, is 'Blessed are those who mourn'.

One of the core reasons modern Christianity has found the message of suffering strange and distasteful is that is has also lost any firm grasp of the nature of Christian hope. It has been my experience pastorally and as a teacher, that the challenge presented by the current climate crisis, and in particular, the increasingly plain failure of humanity to deal with it, has been deeply perplexing to many committed Christians who have given themselves, in hope, to campaigning for change. As the intractability of the political-economic order becomes increasingly clear, the seeming implication is that working for the kingdom has failed – that hope has failed and faith has failed. But this only shows how deeply we have been shaped by Enlightenment optimism about the future. As Lesslie Newbigin and many others have pointed out, much of the modern Church's language about 'building the kingdom of God' has really been a religious gloss to the secular liberal program of Progress.

There is nothing in the New Testament that supports the assumption that the trajectory of secular history is towards improvement, and there is plenty to suggest that things *might* get very bad. Let me stress that I am not here positing the idea present in some forms of millennial right-wing Christianity, that the world is moving towards a prophesied tribulation. Indeed, I am not sure that there is any clear view of the future direction of *secular* history within the New Testament, except to insist that the present age is governed by fallen powers who are capable of great darkness. A quick glance at any era of history bears the truth of this conviction out. If we are thinking solely about *the present age*, then the fullest expression of Christian hope is that found in the opening to the Gospel of John: 'The light shines in the darkness and the darkness *has not overcome it.*' What is clear, and what is so hard for us moderns, is that *ultimate* hope in the Christian testimony lies outside of secular history.

Once again, by long and convoluted courses of history, modern Christianity has either completely lost its grasp of a vital, robust and dynamic hope that stands outside of our present age, or, in those quarters where a concept of eternal hope persists, it has largely become a two-dimensional dualistic and escapist hope that absolves the believer from responsibility for the pain of the world – pie in the sky when you die. This is also reflected in the widely-held (but scripturally untenable) end times theology which looks forward to the rapture and the dissolution of the elements.

In recent times, thinkers such as N.T. Wright have spearheaded a push for the church to recover a more fully biblical worldview, especially in relation to the nature of Christian hope.[5] This is far too big a subject to expound here, but it is worth sketching the contours as it has a critical and direct bearing on how we think and act in relation to climate change:

1. Ultimate hope rests in what *God will do* at the end of the present age (whatever that means) – it is not something we accomplish in secular time, nor is it 'going to heaven when you die'.

2. Christian hope is the hope of resurrection. This is revealed directly in the bodily resurrection of Jesus, which is also promised to all those who belong to him (whatever that means), but takes its fullest expression in a new creation encompassing the whole cosmos, the coming of heaven to earth. That is, Christian hope is radically material, not immaterial, and it concerns not only humans, but the whole of the natural world.

3. Despite a climactic juncture when the present age will end, the life of the future age is not discontinuous with life in the current age. Life that is embraced or denied in this age, is life that is taken up or forfeited in the next.

4. Similarly, but in reverse, through the gift of the Holy Spirit, the resurrection and new creation of the age to come is at all ready at work in those who believe. However imperfectly, the future is an animating force in the present.

There are some who might fear that re-invigorating a sense of eternal hope might encourage escapist denial about the present, or absolve responsibility from acting on climate change, but that is only possible when it is not understood. The sum of all of this, and the underlying point of just about every teaching in the New Testament about future hope, is that the shape of this hope is what directs the shape of our lives and work here and now. What we believe about the future matters critically for how we live now. Revelation provides a final image of heaven coming to earth, but only to a people who are already fully steeped in the prayer of Jesus, 'Your kingdom come, your will be done on earth as it is in heaven'. In Paul's language, Christ's work of reconciling the cosmos

---

5. See especially NT Wright, *Surprised by Hope: Rethinking Heaven, the Resurrection and the Mission of the Church*, Harper Collins, 1989.

is entrusted to us. At the conclusion of 1 Corinthians 15, his fullest and most significant exposition of resurrection faith, Paul succinctly sums up the point of it all: 'Therefore, my beloved, be steadfast, immovable, always excelling *in the work of the Lord*, because you know that in the Lord *your labour is not in vain'*.

Bringing it home

So far, I have been arguing that a fully Christian (by which I mean a fully human and fully fruitful) response to climate change first requires the recovery of certain mental and spiritual perspectives which are central to Christian faith. In particular, we need to become cognisant of the ways in which our hopes and expectations of the future have been profoundly shaped by the secular creed of Progress, and that such hope is both false and ultimately obstructive to positive action. In its place we need to rediscover the New Testament call to both suffering and hope: (i) not only the recognition and explanation of the deep suffering we see in the present age, but also a call to both suffer with, and even to suffer for, the world; and (ii) a clear vision of the end to which all good things work, and how that ultimate purpose directs our life and action here and now.

Given the present state of the church in the West, such a profound recovery of belief seems work enough. But it is not enough, for then there is the question of action – what must we do?

When we talk of a 'Christian response' to climate change, what sort of agency do we have in mind? This is a fundamental question, but one that is not so often clarified. In the little literature that there is, there are generally several levels of Christian agency that are appealed to:

1. The call for individual Christians to become more fully active as political citizens in campaigning for reduction in our collective green house gas emissions.

2. The call for the institutions of the church (which usually means the denominational structures and para-church agencies) to speak out with moral authority about the gravity of our situation and to support the pursuit of serious mitigation measures.

3. The call for local congregations to undertake symbolic and practical actions which increase climate change awareness amongst their membership and within the community.

4. The call for Christian theology to recover a 'Christian political economy' and so begin to live out and to advocate an alternative economic vision from the current model.

All of these dimensions of action deserve strong affirmation.[6] However, on their own, or even all taken together, these forms of action do not represent the half of the Christian calling.

One of the manifestations of the modern crisis of the church is that its public voice and action is just so, well, irrelevant. Any public 'moral authority' which the church may still hold is at best marginal and weak. Furthermore, when the *leadership* of the church do speak with a unified and coherent voice, it can no longer be assumed to represent the majority view of the *membership* of the church, a fact which politicians now understand and exploit all too well. Moreover, when Christians do act politically, whether individually or in groups, it is rare that they have anything distinctively Christian to say, let alone a distinctively Christian *way* of undertaking political action. To put it in both theological and sociological terms, the voice of the church has become a dis-embodied voice – it lacks a body.

How does the Body of Christ re-gain its body? 'You are Christ's body, and individually members of it.' It is a measure of the seriousness of our condition that the implications of this statement have not been clear for some time. We are missing the locus of faith that joins it all together: the thing that joins individual discipleship with the vocation of the Christian body, and that joins ecology and economics with faith. That locus is the household.

It is no accident that the three words 'economy', 'ecology' and 'ecumenical' all share the same Greek root: *oikos* meaning 'house'. Economics, a term which has been perversely abstracted and mystified,

6. Most of them are brought together in some way in the Hope For Creation Campaign. See http://hopeforcreation.com.au

refers to the management of the household: the work, production, distribution and consumption by which all households must somehow live. Ecology refers to our understanding of the great web of life which makes up the household of nature. Wendell Berry writes of the ecological order as 'the Great Economy', and refers to human economic systems as 'the little economy' which exists within and depends upon the greater. Ecumenism is the movement which seeks unity within the household of God, the Church Universal.

None of these three – economics, ecology and the church – can be properly understood without understanding what is happening in actual households. The household is the foundational locus of work, consumption, leisure, relationship and community; it is the foundational locus of our use and mis-use of the Earth; and the household, not the congregation, is the primary place where faith is enacted or denied in a thousand choices every day.[7]

It should not surprise us then, that the conduct of households is a central and continual theme in the great Biblical narrative of salvation. To put it another way, we can say that at the heart of the vocation given to the people of God from Genesis 12 through to Revelation 22 is the task of embodying the reconciliation of economics, ecology and faith in their everyday lives. We cannot properly understand Israel, the Kingdom of God or the Body of Christ without understanding this dimension of their meaning. The two big words which are used to describe the shape of the lives that the people of God are called to live are 'holiness' and 'shalom'. Holiness shares its root meaning with 'wholeness' and 'health': it describes the integration of the life of body and spirit, the undividedness and the 'health' which is God's. Shalom refers to all things in right relationship – God, humanity and creation.

These two things, holiness and shalom, can be said to be the organising principles of Biblical household economics and they reverberate throughout the Books of the Law, the prophets, the teachings of Jesus

---

7. I am, in this statement, including the workplace as an extension of the household economy, where it belongs, rather than something separate and divorced from the household.

and the pastoral instruction of Paul.[8] Throughout they require constant translation into the practical realities of everyday life, a process which is therefore fluid and dependent on context. However, the hallmarks of holiness and shalom in everyday life are consistent throughout the Bible, and a fair summation of them would be something like:

1. The rejection of false gods.
2. Justice and neighbourliness.
3. Care and nurture.
4. Lifting up the lowly.
5. Abundant life.
6. Light for the world.

What might these hallmarks look like in the household economics of twenty-first century Australia? This is a vast subject that requires elaboration in a thousand different directions, but let me suggest just one foundational implication that relates directly to the challenges of climate change. It drives right to the heart of the Gordian Knot of climate change that no political party in Australia, and few environmental groups, have been willing to sever; it also drives straight to the core infidelity of much modern Christianity, our transgression of the First Commandment: 'You shall have no other gods before me.' It is simply this: we must accept a reduction in our material standard of living. More than that, we must *embrace* a reduction in our material standard of living.

One of the reasons the modest carbon reduction initiatives of the Rudd and Gillard Governments were so quickly undone (and why they were so modest in the first place) is because it is so easy for opponents of such changes to show that they must inevitably result in an increased cost of living and therefore a decline in disposable income. The price of electricity has occupied the frontline of this attack, but climate change opponents

---

8. I have written on this idea at length in J. Cornford, 'Longing for a better country: Christianity and the vocation of social change', *Zadok Paper*, Winter 2007, and in numerous articles in *Manna Matters*, collected here: <www.mannagum.org.au/faith_and_economy/economics-bible>

have quite rightly demonstrated that the increased cost of living must flow through to all manner of other things. And as long as continually rising material standards of living remain the golden calf before which we must all bow, opponents of climate change are on a winner.

For their part, the green lobby have desperately tried to argue that the economic advantages of a shift to a 'green economy' would offset the costs of moving away from fossil fuels. There is undoubtedly truth in this, but no one really believes that the staggering extractive wealth of the Oil and Coal Age can be matched by renewable energy. The elephant in the room must be faced – we can no longer go on living the way we are. Things must change:

- We need to use less energy and pay a higher price for non-polluting power generation. Reductions in energy use can be achieved to some extent by increased energy efficiency (which will mostly have to be paid for), but it must also come through less extravagant use of cooling and heating, something which increased prices will no doubt take care of.
- We need to pay a higher cost for transport fuels (especially aviation fuels) and make far greater use of non-oil transport. This means we will need to travel less ourselves (especially by air, but also by car) and we will need to pay higher transport prices in all the goods we buy, especially food.
- We need to reduce both organic waste (which produces methane) and inorganic waste (which requires replacement mining, manufacture and transport) going to landfill, which probably requires paying a much higher price for waste services.
- We will increasingly need to source more of our food, and perhaps industrial production, closer to home, which means we will need to cover higher wage costs and regulatory standards in the things we buy.
- All of the above means that food and industrial production will become much more expensive, which means that the per-unit cost of everything we buy will become higher.

- All the while, the economic adjustments that all of this sets in motion will render many currently profitable enterprises unviable. Employment insecurity will likely increase and real wages will likely be driven down.

- All of this adds up to this: higher cost of living + lower incomes = lower material standard of living.

[I have here, for arguments sake, only outlined some of the changes directly required by climate change; however, if we were to consider the other global ecological crises of species extinction, resource decline, water scarcity and soil depletion (which we must), many more changes would need to be listed.]

The scenarios which I have outlined above currently constitute political heresy of the worst kind. The fact that they will likely happen anyway, irrespective of whether we choose them or not, is neither here nor there. Such truths cannot be spoken, and to do so requires the speakers to be thrown into the hottest Babylonian furnace of political derision that can be collectively stoked by political, commercial and media elites. Of course, to be thrown into such furnaces is precisely the vocation of the people of God.

In a world where the prospect of a reduction in standards of living seems like a kind of death and the end of the world, it is the vocation of the church to show that it is actually the path to life and the saving of the world. The thing that is being denied and avoided is the very thing that the people of God *choose*. The false god of Western affluence must be shown for what it is – a lie that promises life and delivers death; a lie that requires the sacrifice of our children.

All of the required changes listed above, even lower incomes, *can be chosen now*. Indeed there are people who are already making these choices. What does it look like practically? I will not elaborate in detail as there are so many resources which do this well, but the contours are simple:

- Paying higher upfront costs (lower costs in the long term) for energy efficient homes (insulation, double glazing, draft sealing, thermal mass, seasonal shading, better appliances etc).

- Paying higher costs for renewable electricity (Green Power options) and/or upfront costs for solar panels.
- Choosing organic, more local, and more ethical food, which costs more.
- Maintaining, repairing and re-using manufactured goods, rather than disposing and replacing, which costs more in time and money. In particular, resisting the drive to continually update and replace technology.
- Growing food and processing food (bottling, preserving, jams and spreads, cordials, bread, pasta etc) at home and in local groups.
- Closing the loop on organic waste through intensive composting going towards food production, so contributing to absorption of greenhouse gases rather than emission.
- Living generally on a smaller geographic scale, not just in sourcing of consumer goods, but through the limiting of travel, especially air travel, and staying to a greater extent within human-powered limits (walking and riding) or public transport limits.
- At the household level, work less hours in paid employment (more equitably shared between men and women) to invest more time in the work and production of the household economy (more equitably shared between men and women). Such a lifestyle may well be less compatible with certain career aspirations (and the associated income advancements).

This is the reconciliation between economics and ecology at the household level that the earth requires of us. But lo! Behold! It is also the place where the ideas of faith are enfleshed in lived experience. Because the very things the planet requires of us are the very things which make for a better, deeper and richer life, full of stronger connections within households, between households, and with the earth itself – a life in which the Creator who 'is all, in all' can be more fully known and worshipped. It is the way that we will more fully come to understand the truths at the core of Christian faith: that by seeking to save your life you lose it, but in losing your life you will find it; that the meek inherit the earth; that we do not live by bread alone; that we are all called to join Christ's work of reconciling the cosmos to himself. And the more that lives (households) are conformed to such truths, the more they will be able to seen by the world at large, and the less hollow they will sound when spoken. In placing household economy at the very centre of faith, not only will Christianity have reclaimed the material substance of its spiritual message, it will also have re-discovered its primary evangelical resource and the basis of its prophetic voice.

At this point, the person who is well acquainted with the challenges posed by climate change may offer something of an objection: 'How can you say that this is the vocation of Christians? These are the very changes that will be required of all humanity if we are to be saved.' Precisely. And that, in a nutshell, is the vocation of the household of God: to be the first fruits of a new humanity; to live according to the re-creation of all things now. That is why to be fully Christian is merely to be fully human.

In concluding, let me clarify an important point. I am not suggesting that if Christians make these changes in their lives and households, they will then be transformed into 'effective' advocates for climate change mitigation policy which well then finally set the ball rolling and the world will be saved. I am not arguing for renewal of faithful Christian households because it is what advocates and lobbyists would call 'an effective model of change'. If there is some sort of significant renewal of Christian belief and practice along the lines I am suggesting, then it might well make the church a more authoritative voice with more influence in social and political change. On the other hand, it may contribute to an even more aggressive marginalisation of the church and a sharpening of conflict between secular culture and the Christian message. The nadir of human failure on climate change may yet be worse than the current moment. All of our best efforts may yet fail. Who can say?

I am arguing that the household of God is called to embody within itself the reconciling of economics and ecology *irrespective* of whether we are 'successful' in convincing others to do likewise. We are called to dedicate our lives as 'living sacrifices' out of love for world, because our salvation is bound up with it, and simply out of obedience. If we can do so, no matter how dark the times may become, there will always be a light that shines in the darkness. We have no theological assurances at all about how secular history will proceed to unfold in this century, but we have a whole Biblical cannon that attests that the fulcrum of human life, of all life, is secure, whatever may happen. My prayer is that my daughters will grow up in a broader household of faith where there is a deep agreement with the confession of Julian of Norwich that I began with: 'All shall be well, and shall be well, and all manner of things shall be well'. And may we also share the same assurance of the said saint: 'The worst has already happened and been repaired'.

LIVING IN BABYLON
Finding shalom in the city

Babylon,
Born in your walls
Bred in your will
Captive, but still ...
I hear the heavens cry ...
Babylon,
This is goodbye.

Steve Taylor,
I Predict 1990

When driving to Melbourne from Bendigo, there is a point where the freeway, as it descends down the range, sweeps around and gives a first sight of the city in the distance. It is an impressive sight. Out of the midst of a great plain rise the towers of the city, dominating the landscape. On a foggy day, it can even look as though the towers rise out from the clouds. The sight still grips me with a mix of awe and horror. And most times, a single thought impresses itself on my mind: Babylon!

Cities are incredible achievements of human ingenuity and they are amazing places to be. When I am in Melbourne's city-centre my pulse always runs a bit quicker. The scale of the human project on display in the city is just so impressive, and, well, awesome. The presence of so much money and power, and of so much activity and ingenuity, cannot leave you unaffected.

There is a lot of superficial talk about cities. Probably the dominant voice is the one that proclaims how wonderful cities are as places of culture, creativity, possibility and freedom. Of course, there is much truth in all this, however, its celebratory key also serves to obscure deeper truths about the city. At the other extreme, less common but perhaps getting stronger, is the characterisation of the city as a place of evil, of untold damage – a place we must get away from. Once again, there is much truth in this story, but on its own, this story offers little hope or guidance.

Most people reading this live in a city, and those who do not are reading it by virtue of developments in organisation and technology only made possible by the city. As Jacques Ellul says in his seminal work, *The Meaning of the City*: 'The city is everywhere.'[1] However, there has perhaps never been a time in human history when we have more needed to think clearly about the city: about what it is; about how we live in it; and about what it could be. If we are interested to uncover the ways in which our economic lives affect the global poor and the health of the planet, then we cannot do this without thinking hard about the city. And if this task is critical for humanity then, of course, it is critical to the concerns of the Christian gospel. So it should not come as any surprise that the Bible has something essential to say about cities. Essential, because we need to hear it, but also essential in that what the Bible illuminates is the very essence of the city. But before turning to the Biblical story, it is worth putting the city of today more clearly into focus.

An urban species

In 2008, very quietly and with little fanfare, a momentous point in the human story was reached: we became an urban species. In that year, we reached the point where more than half of humanity lived in cities. By 2050, it is estimated that about 75% of the world's population will live in cities, and about one-third of these city dwellers will live in slums.

The face of the globe has been irreversibly altered by cities. The planetary crisis now facing us – the changing climate, species extinction, the loss of arable land, resource depletion, the plundering of the oceans – has its source in the city. Even the very considerable impact of agriculture upon the earth is, by and large, the product of the types of agriculture required to supply cities. Similarly, wherever it is found, the existence of poverty and inequality cannot be properly understood without understanding the role of cities in shaping all human relations.

Indeed, we may simply say that most products of human civilisation, whether glorious or appalling are at heart, the products of the city.

---

1. Jacque Ellul, *The Meaning of the City*, Paternoster Press, 1970.

The word 'civilisation' derives from the Latin civitas, meaning city, or city-state. The very beginnings of what historians call 'civilisation' in the Fertile Crescent (modern day Iraq through to southern Turkey and northern Egypt) of Neolithic times, is really the beginnings of the rise of the city in human affairs. At the heart of this project were a number of developments in Fertile Crescent urban settlements that remain foundational to urban civilisation, although they are somewhat obscured from our eyes today:

1. the development of an organised military enabling social control and coercion, protection from external threats, and the ability to expand and dominate new territories;

2. the reorganisation of agriculture through large-scale, centralised infrastructure (primarily irrigation) producing greater surpluses and greater elite control over surpluses;

3. the refinement of royal, city-based 'state religion' that legitimated power structures and provided the social glue which is everywhere necessary to ensure the survival of any complex system;

4. the rise of trade between urban centres, each trading goods exploited from their own hinterland;

5. and finally, last in the historical process was the development of writing, which enabled the more efficient administration of such a complex social, economic, political and religious system.[2]

From this early development of a few city-states amongst the vast morass of unorganised and 'uncivilised' human communities, cities have gone on to exert their dominance over the Earth's entire landmass, its oceans, and even the regions of space surrounding us. One way of looking at a map of the world is not to see a division of nation-states, separate and competing, but to see a global network of urban centres exerting their dominance and control over the resources of the planet. Indeed, modern cities have now come to assume a scale that adds new

---

2. It is worth noting that while we are here focussed on characteristics and developments that are central to the city, we should not make the mistake of assuming that they are exclusive to the city.

dimensions to the city (there are now a number of mega-cities around the world with populations greater than the Australian continent):

- they are places that can no longer live off their hinterland, but must draw resources from the planet;
- they are places where any broader cohesive community is impossible;
- and, more controversially, they are places where ugliness has gained a final triumph over beauty.

So far, this description of the rise of cities seems to paint a dark picture. However, it is from this very basis that humanity has also attained all those accomplishments that most of us would want to celebrate: accomplishments of art and architecture; science, medicine and engineering; literature and philosophy; and, ever so slowly, the development of new ideas about the structure of society, economy and government.

This is where there is often so much difficulty in understanding human affairs; when things we like and value are so closely connected to things we dislike and disown, it generally seems easier to either ignore or deny the darker elements (the status quo option) or to ultimately reject the tainted good things (the purist option), rather than to try to see and hold together the whole messy story. And it is indeed the whole messy story about the city that is presented us in the Bible.

The meaning of the city

It is common to refer to the story of Adam and Eve and the forbidden fruit in Genesis 3 as 'the Fall', identifying this as the point at which Eden is lost. It is probably truer to say that the Biblical narrative of 'The Fall' begins in chapter 3 and extends to the end of chapter 11. 'The Fall' account of Genesis is indeed a succession of stories that lay bare the human condition, and the city occupies a prominent place in this story.

In the Genesis account of primeval history, the city has its origin with Cain, the killer of Abel, the first murderer. The progression of the story is significant: when Cain kills his brother his act of violence simultaneously upsets the ecological order and unleashes a state of permanent insecurity upon himself and all his descendants (Gen 4:10-12). Unrepentant, Cain's defiance continues: he turns away from the presence of God and enters the land of Nod – the land of wandering.

It is from this place, the place of wandering insecurity in a wounded creation, devoid of God, that Cain begins his own project for permanency and security – he fathers a child and builds a city. Both the child and the city are given the same name, Enoch, which means 'initiation' or 'dedication'. With Eden lost, Cain is inaugurating the new project that we have come to call 'civilisation' – city building. To quote Ellul: 'For God's Eden he substitutes his own; for the goal given to his life by God, he substitutes a goal given to himself.'[3] In this story of origins, the city is a response to the insecurity and ecological uncertainty that man has unleashed with his own violence, but even more importantly, the city is the manifestation of humanity's attempt to live independently from God.

From this point, the Genesis narrative recounts the plunge of humanity into ever more violence and discord, until 'the whole earth was filled with violence' (6:11). After the Noah story, the city comes into prominence again with the activities of Nimrod, the first 'mighty warrior' (10:8). Most translations rather oddly cast the next description of Nimrod in a rather positive light, 'He was a mighty hunter before the Lord' (10:9). The sense of the meaning is much more likely negative: he was before the Lord in the same way an accused criminal stands before a judge. In other words, in God's judgement this 'mighty warrior' (and note, so far in Genesis, violence has only been seen as a sign of distance from God) was essentially a hunter; but what did this hunting look like? The next line is striking: it tells us that Nimrod was a city-builder – the city is something that exists by preying upon its surrounds. It is no surprise then that the list of cities attributed to Nimrod is a roll-call of the cities that will form the Babylonian and Assyrian empires later responsible for the cataclysmic overthrow and exile of God's people. In these very concise verses, what is being described is the next stage of the human

---

3. Ellul, *The Meaning of the City*, p.5.

project: domination, subjugation and exploitation. And according to the Genesis account, the city stands at the heart of it all.

Genesis' primeval history of the city does not end there though; it has one more dramatic tale to tell, which brings the narrative of the Fall of humanity to its climax – it is, of course, the tale of Babel.

The confusion of the city

To put this story in context, there is now a fair amount of scholarly consensus that the final form of the stories we have received in the first five books of the Bible (the Books of the Law, or the Torah) was completed either during, or soon after the period of Hebrew exile in Babylon in the 6th century BC. Thus, to gain the full power of the tale of the Tower of Babel in Genesis 11, we need to imagine in the background the world's largest city at the time – the living, breathing, all-dominating city of Babylon – itself dominated by the great edifice of the real, historical Tower of Babel, the Great Ziggurat of Babylon.

In the Babylonian tongue, 'Babylon' means 'Gate of the Gods', and in their own creation story, Babylon is created after the violent and gory triumph of Marduk as ruler of the Gods. Babylon was Marduk's seat, and the Babylonian king his representative on earth. In the shadow of this awesome city-empire, the Hebrew exiles write a very different account of what is going on.

Strikingly, in the Genesis account, the building of the city and the tower is a project of all of humanity (11:1-2) – everyone is implicated in this story. The story goes on to describe how people agree to 'build a city, and a tower with its top in the heavens'. As 'the heavens' in the ancient world was where God was thought to reside, this was from the outset a challenge to the supremacy of God. But even more significantly, the story records the people's deeper purpose: 'let us make a name for ourselves'. This is not, as it would seem in our idiom, a reference to becoming famous, but something more profound. In the ancient world a name was bestowed upon someone by their progenitor, and in the Bible a person's name often says something essential about their character.

Here, humanity wants to make a name for itself – that is, humanity is rejecting its creator and attempting to engage in an act of self-creation.

There is perhaps no better way of capturing the spirit of modern (and post-modern) humanity, than to describe it as hell-bent on the project of self-creation. This is evident in the ways that identity has become endlessly fluid and increasingly staked to consumer fashions; it is evident in the use of social media to present a constructed version of yourself to the world; it is evident in the belief that modern science can and should embark upon manipulation of the human genome to 'improve' the species in the ways that 'we' desire.

But while humanity is endlessly impressed with its own project, the Genesis tale highlights the hubris and irony of it all. Although to the people, the tower seems to rise to the heavens, God in the story has to 'come down' even to see it (v.5). And in the end, the whole project – like so many bad building jobs – is scuttled by the breakdown of communication. But, of course, the point is deeper than that – the human project of independence from God and self-creation is ultimately undermined by the inability of people to construct a truth that is valid for all. The city that called itself the Gate of the Gods (*Babili*) is renamed in Genesis as Babel, meaning 'confused'.

And with the story of Babel – the confused city – so ends the Genesis account of 'The Fall' of humanity. City-building is certainly not the cause of this fall, but it is a prominent manifestation of it. In essence, these ancient stories from Genesis are telling us what we should already know from our day-to-day lives in cities: whatever the profession of our faith, the city tends to render all of us practical atheists. The city is the outworking of all our attempts to live independently from God and nature, and to fashion life in this world according to our own purposes. To quote Jacques Ellul once more: 'Just as Jesus Christ is God's greatest work, so we can say, with all the consequences of such a statement, that the city is man's greatest work.'[4]

---

4. Ellul, *The Meaning of the City*, p.154.

God enters the city

One might think that with such an introduction, the biblical stance towards the city is therefore one of rejection and opposition. Not so. How surprising that after concluding the story of the Fall with Babel, the story of hope and salvation for humanity moves inexorably toward another city: Jerusalem!

The story of Jerusalem in the Bible is perhaps the hardest of all to tell. It is so full of contradiction: a story of grace and judgement; of unutterable despair and of soaring hope. Jerusalem's name means 'City of Shalom', but has there ever been a city in human history that is such a lightning rod for conflict? And perhaps it is the starkness of the contradiction between the vision of Jerusalem as the Holy City and the earthly reality of her plight that makes Jerusalem so significant. At its deepest level, Jerusalem is a sacrament: that is, Jerusalem reveals the truth of every city, the whole horrible and wonderful truth.

The Second Book of Samuel (5:6-12) tells us that David chooses and conquers Jerusalem to be his new capital after consolidating his kingship over Israel and Judah. The subsequent secular history of Jerusalem seems to confirm everything that Genesis has instructed us to expect from the city: the growing domination of people and the land, economic exploitation, excess, violence and idolatry. The prophets are full of lamentations waxing lyrical about the depths to which Jerusalem sinks: 'How the faithful city has become a whore! She that was full of justice, righteousness lodged in her – but now murderers!' (Isa 1:21). Indeed, the prophets are full of condemnations of cities – Babylon, Nineveh, Tyre, Sodom – but all of these pale compared to the condemnation that ultimately falls to Jerusalem. The prophet Ezekiel proclaims that next to Jerusalem, Sodom and Gomorrah appear righteous (Eze 16:43-52). So at one level we can say that in the Biblical narrative, far from enjoying any moral superiority, Jerusalem typifies the character of the city, and it is not a pretty picture.

But with David's choice of Jerusalem, something else truly startling happens. For we are told repeatedly that despite all her subsequent history, God chooses Jerusalem. (Eg. Deut 12:4-7; Eze 16: 1-14). Although, as a city, Jerusalem represents the culmination of man's choice, man's work and man's rebellion, God nevertheless says 'yes'. God comes into the city. In a move that pre-figures the incomprehensible grace of God's movement towards us in Christ, God enters into the heart of human fallenness and says 'I will make you Holy'. And so, confusingly, parallel to all those texts denouncing the sin of Jerusalem, the Hebrew scripture also contains a host of texts extolling her wonders and delights. And in the prophets there is a new note, a new vision of the city: the city as she could be; the city as she ought to be; the city as a place where man's work has been joined to God's:

> For out of Zion shall go forth instruction [*torah*], and the word of the Lord from Jerusalem. He shall judge between many peoples, and shall arbitrate between strong nations far away; they shall beat their swords into plowshares, and their spears into pruning hooks; nation shall not lift up sword against nation, neither shall they learn war any more; but they shall all sit under their own vines and under their own fig trees, and no one shall make them afraid; for the mouth of the LORD of hosts has spoken. (Micah 4:2-4)

This is a vision of the city in harmony with the country-side, a place of peace and justice, a place of shalom. And ultimately the salvation and redemption of Jerusalem is the salvation of every city. Jerusalem is every city.

These parallel, seemingly contradictory, attitudes to Jerusalem are continued in condensed form in the New Testament. Perhaps nowhere is God's grace and the tragedy of the city more poignantly captured than in Jesus' own confrontation with Jerusalem:

> But as they came closer to Jerusalem [City of Shalom] and Jesus saw the city ahead, he began to weep. "How I wish today that you of all people would understand the way to peace [shalom]. But now

it is too late, and peace [shalom] is hidden from your eyes." Luke 19:41-42)

"O Jerusalem, Jerusalem, the city that kills the prophets and stones God's messengers! How often I have wanted to gather your children together as a hen protects her chicks beneath her wings, but you wouldn't let me. And now, look, your house is abandoned and desolate." (Matt 23:37-38)

As we know, the whole dramatic tension of the gospels is built around Jesus' journey to Jerusalem, the place where he will die, hanging on a gibbet outside the city.

Despite God's reaching out to her, the Jerusalem of Jeremiah's time and the Jerusalem of Jesus' time are both brutally destroyed (by Babylon in 587 BC and by Rome in 70 AD). It almost seems – and this is reinforced by Jerusalem's history to the present day – that if we are to see Jerusalem as a sign of hope for the city, then we must also be confronted with its history as a sacrament of suffering. Perhaps we cannot see one clearly without the other.

But, of course, the final pages of Christian scripture do end in hope. More than that, they provide a ringing, climactic, endorsement of the possibility of the city. The city may be man's greatest work, but God has joined with that work and brought it to a completion that man could never envisage on his own: 'And I saw the holy city, the new Jerusalem, coming down out of heaven from God' (Rev 21:2). It is in this ultimate city that God makes his home among mortals (v.3), a city where human suffering has ended (v.4), and a city of ecological harmony, where both the river of the water of life and the tree of life bring healing to the nations (22:2-4).

Beautiful. But what is that to us? That is a vision of what is to be – it is not where we live now. It is given in the concluding chapters of the Book of Revelation as a joyous, beautiful and inspiring confirmation of the end to which all good things work, but it is not where John's letter has its rhetorical punch. That comes a few chapters earlier, when this letter to the churches reveals to us the place in which we are living right now. We are living in Babylon.

Exiles in Babylon

It is not too much to say that the shadow of Babylon hangs over the whole Bible. Like Shadrach, Meshach and Abednego, the Hebrew scripture as it comes to us was largely refined and finalised in the furnace of the Babylonian exile. The stories of humanity's fall, of Abram's call, of liberation from Egypt, all bear echoes of the later exile in Babylon. Indeed, it is not too much to say that these stories represent a refutation of Babylon and all that it represents.

For the Hebrew people, the destruction of Jerusalem and the exile to Babylon is a crisis like no other. It is what the crucifixion of Jesus was for the disciples – the seeming utter defeat and trampling of all that they had come to hope for and see as good in the world. It seemed to be the very overthrow of the God they had worshipped.

For this reason, the image of Babylon – the Great City – has in the Bible assumed a special significance as the archetypal enemy of God. Actually, viewed with some historical objectivity, Babylon was no worse, and a probably a good deal better, than many other city-state empires that have arisen over the aeons. But because of its particular relationship to Israel, the name Babylon comes to be a potent symbol of power, violence, wealth, oppression, greed, excess and the culmination of human arrogance and rebellion. Wherever in the Bible it is mentioned, or even echoed, we should pay attention.

For this reason, it is hard to over-emphasise the significance of the appearance of Babylon in the final pages of the New Testament. Contrary to much supposition, the purpose of the Book of Revelation is not to give us a Nostradamus-like prediction of how the world will end, but rather to lift the veil on reality and show us clearly the world we inhabit now. It is in the form of a letter to the churches, calling them to see the world from a God's-eye perspective and to stay faithful to the good news they have received. It is addressed to churches in seven cities of

Asia Minor (modern day Turkey), then a very prosperous, comfortable, secure, loyal and highly urbanised corner of the Roman Empire. It used to be thought that Revelation was written during a time of persecution of Christians, but that no longer seems so certain; rather, the internal evidence of the letter suggests that the churches in these seven cities occupied varying social positions, some marginal, others comfortable and prosperous (see chapters 2-4). Whatever the case, they all lived in a region where belief in the beneficence of Roman Imperial rule was deeply held.

How significant then that the letter reaches its climax with the angels crying out to the churches:

> Fallen, fallen is Babylon the great! … Come out of her, my people, so that you do not take part in her sins, and so that you do not share in her plagues; for her sins are heaped high as heaven (Rev 18:2-5).

Here the churches are told that their 'relaxed and comfortable' lives were being lived in the very midst of the enemy of God – Babylon the Great! The sins of Babylon (and the Beast with whom she is associated) are listed in explicit detail: she is a dominating power; she demands worship and allegiance; she 'blasphemes' God – that is, she takes God's position to herself; she seduces the earth with her wealth and luxury; she deceives the earth with her amazing wonders; she oppresses the poor; she enslaves the very human soul; and she undergirds all of this with violence (see chapters 13, 17 & 18). Ring any bells?

And the implication of John's letter is that, somehow, God's people have become party to these sins! Ring any more bells? Of course! It is Babylon that we have been living in all along. When we rip back the veneer of our progressive liberal democracy we can see that we are in fact living in a global network of urban humanity that has been plundering the earth, enforcing its will with violence and caring little for the human toll. And the people who today are looking clearly and unsentimentally at this system are indeed crying 'Fallen, Fallen is

Babylon the great ... do not take part in her sins so that you do not share in her plagues!'

Indeed, if the message of the Book of Revelation had to be boiled down to one quote, then it would likely have to be this: 'Come out of her, my people, so that you do not take part in her sins!'

But how? How do we depart from Babylon? What is clear in Revelation is that Babylon is no longer just a place, and it is even more than a system; it is the towering project of humanity challenging the heavens, seeking to be the authors of our own creation. How do you leave such a thing? Babylon is everywhere.

While the Book of Revelation is effective shock treatment to wake us out of our apathy, it is not a very practical book. All that titanic struggle between light and dark is not much like everyday life. Interestingly, perhaps one of the most useful Biblical texts for helping us come to terms with our situation is the letter of the prophet Jeremiah to the actual exiles in the real, historic city of Babylon (Jer 29). Here, the prophet is writing to a people in the midst of that crisis of faith discussed above: 'By the rivers of Babylon, we wept when we remembered Zion ... How can we sing the Lord's song in a strange land?' (Ps 137). They are trapped in a place not only foreign to them, but entirely hostile to their faith. Put simply, they are not at home – they do not belong. But, startlingly, the prophet's advice seems, superficially at least, to be almost the opposite of that given in Revelation:

> Thus says the LORD of hosts, the God of Israel, to all the exiles whom I have sent into exile from Jerusalem to Babylon: Build houses and live in them; plant gardens and eat what they produce. Take wives and have sons and daughters; take wives for your sons, and give your daughters in marriage, that they may bear sons and daughters; multiply there, and do not decrease. (Jer 29:4-6)

Of course, in Revelation we are called to depart from the spirit of Babylon, not a physical place, while Jeremiah is talking to people who do not have much choice about their physical location. Rather than plunge into existential angst, he advises them to do something so simple that its profundity can easily be missed: live life and live it well. The things he calls to attention – homes, gardens, food, love, children, community – evoke the simplest and the fullest blessings this life has to offer. To those in the midst of Babylon, the great destroyer, is given the call to nurture whatever is good, whatever is wholesome.

But in case we haven't caught the full significance of this, Jeremiah goes on: 'Seek the shalom of the city where I have sent you into exile, and pray to the LORD on its behalf, for in its shalom you will find your shalom' (v.7). Even as exiles in a city foreign to their faith, they are instructed to live for, to work for, and to pray for the shalom of the city – everything in right relationship, even Babylon the Great.

Although the period of exile was, in historical terms, not all that long, it has come to be a defining motif in the Bible, and especially so in the New Testament. When the letters of Paul, Hebrews, Peter and John are read with this idea front and centre, they take on a whole new light, as do the teachings of Jesus about being salt, light and leaven in the world. And for those trying to follow the way of Jesus in 21st century urban consumer culture, this is precisely our position. We are exiles.

Living in Babylon

However we might conceive of the ideal life, the reality is that most of us are bound to the city in some way. As humanity moves deeper and deeper into the urban world in the coming century, more than ever we need people who can re-conceive what it might mean to live well in cities. If we take the Biblical testimony seriously, than this must require a deep recognition of the spirit and essence of the city and a preparedness to engage in a relentless spiritual struggle to resist both the seduction and the intimidation of Babylon.

But the struggle is also over the minutiae of our daily lives. Put simply, we are enjoined by the Biblical witness to stop being a part of the problem and to start being a part of the solution. And that requires living differently from the norm.

What does it mean for us, in our time, to heed the call to depart from the sins of Babylon? What might it look like to follow Jeremiah's instruction to live abundant lives, while working for the shalom of the city? How do we, like Abraham, seek what it means to live faithfully in the world as we find it now, all the while 'looking forward to the city that has foundations, whose architect and builder is God' (Heb 11:10)? In short, how do we live well in cities?

I shan't pretend that I can answer this enormous and endlessly complex question. Our life circumstances are all so different and context is so important. Nevertheless, I would like to modestly suggest four areas of our lives in cities which at least deserve serious consideration, whatever we might feel our scope for action is in those areas.

1. Recognising the city as a spiritual force

I was once talking to a young man from Melbourne who now lives in Bendigo. He was saying that although he was a part of groups and communities that talked a lot about the pressures of consumerism, it was not until he had spent some time living outside of Melbourne and then returned to it that he became fully aware of just how monumental is the onslaught of advertising and social pressure to participate in consumer hedonism. (These things are all present in Bendigo too, just not on the same scale.)

There is a huge literature that describes sociologically and psychologically the various effects that cities have on us individually and as a society. But the Biblical perspective takes us one step further, for it recognises the city is a spiritual force. And at the heart of the spirit of the city is the continual push towards disconnection. The primary manifestation of this disconnection, but perhaps the least noticed, is the way that the city provides the illusion of living independently from God. When continually surrounded by such an intense, complete and (seemingly) well-functioning human system, God becomes increasingly obscure, abstract, and for many, irrelevant. Whatever our religious confession, the city encourages us into practical atheism.

Disconnection from God leads inevitably to disconnection from one's neighbours, and this is an effect of the city that most people are aware of at some level. The more humans you crowd into a small space, the less available they become to each other as human beings – think of a packed peak-hour commuter train, where thousands of people are crammed into a highly intimate proximity, all the while studiously ignoring each other's existence. This is an understandable psychological defence mechanism, however the habit of ignoring the realities of other people is bad for our souls and bad for our societies.

Of course, by definition, the city is also a place of disconnection from creation. The implications of this are far greater than being distant from quiet places in nature where we can indulge in reflective sighs. The true cost of this disconnection is that we just cannot see one of the foundational truths about ourselves, which is that we are still entirely dependent upon the bounty and beneficence of the natural economy, and our inability to see this truth is killing us and the planet.

Although these disconnections are profound, we are either not conscious of them or apathetic about them, because the other genius of the city is its ability to seduce, divert and distract. Movement, lights, energy, grandeur, scale, food, culture, sub-culture, shops, opportunity, freedom, choice, anonymity, power, money, success – taken together all of these things amount to a very deep temptation to become complicit in the disconnections at the heart of the city.

If these things are true about the spirit of the city, then the Christian vocation in the midst of the city should be clear: it is re-connection. As with all false spirits, the spirit of the city must be resisted, which requires a reconfiguring of our mental world, and daily practical acts of defiance and resistance. We must find the ways to re-connect with God, with each other, and with the land, and we must find ways of resisting the seductions and distractions of the city.

2. Re-connection with people and place

One of the major forces for disconnection in the modern city is simply the way in which the geography of our lives is divided up over large, hard-to-travel, spaces. Home, work, friends, school, church, and shopping tend to be geographically isolated from one another and we effectively become placeless people. This has effects upon our stress levels, upon the planet, and severely limits what sorts of things are possible between people.

Therefore one very practical measure to counteract the disconnections of the city is to try, as much as possible, to limit the geographic scope of our lives and live as much as we can within our localities. I hasten to add that this is not something that is easily achieved, however we have no hope of improving the situation if we do not at least become more seriously conscious of the cost of spreading our lives across the city, and when we get the chance, starting to make alternative choices.

From a Christian perspective, this probably means that we need to make efforts to re-localise the church. The trend towards larger churches has tended to abstract the experience of Christian community from the experience of place – that is, from the experience of day-to-day life. It also means a church tends to be disconnected from the life and concerns of the immediate local area, and especially the concerns of the marginal and hurting. Re-localising church could mean a re-invigoration of smaller local churches, or it could mean larger churches putting much more energy into establishing localised cells and groups. When Christian groups begin to share something more substantive than just a statement of belief – schools, parks, shops, good works etc. – then it is much more possible for them to begin to apply their faith to these contexts.

3. Re-connection with the country

Which brings me to another critical re-connection. One of the most socially and ecologically damaging effects of urbanisation has been the way it has increasingly driven a wedge between the interests of the country and the interests of the city. City people tend to become dismissive and contemptuous of the country, and country people learn to be suspicious and hostile to the city.

One of the most practical and urgent things for city dwellers to begin reconnecting with, is food and the way we eat. Beginning to explore this area opens up how we connect with our neighbourhoods, with the people in them, but also with the country, with food producers, and with the earth itself. This requires becoming much more familiar with the practices and habits of 'ethical consumption' that have been so helpfully elaborated in recent times, beginning to be more discerning about what food is purchased, where it is grown, how and by whom, and through what channels it is purchased. Not so long ago, this was an insurmountable obstacle to ask of consumers; now the information, frameworks and pathways are well at hand.[5] But it should also involve the city beginning to shoulder some more of the burden of raising its own food – the exciting field of urban agriculture has only barely begun to open up in recent times. At a fundamental level, the city must become more aware of its dependence on the country, more grateful for what it produces, and more responsible for ensuring that the prices we pay, and the ways we get food, enable people, animals and the land to be treated with reverence and respect.

4. The challenge of housing

Perhaps the biggest and hardest challenge to living well in contemporary cities is the challenge of housing. One of the great travesties of our wealth-driven market economy is that the perceived interests of so many have been staked to driving up real estate prices, making the fundamental human need for housing one of the biggest arenas for profit-taking. The price of housing, whether ownership or renting, is an underlying structural determinant of what other life choices become possible for us in location, work, community etc. It determines the incomes we need to somehow raise and how much time we have available for other things; it affects our ability to locate ourselves more closely to friends and/or family; it limits how much money we can spend on making our homes

---

5. See www.ethical.org.au to get started.

more sustainable and efficient; it affects the nature and possibilities for church communities.

If we are to find ways of living well in cities – with deeper and fuller connections – then we need to find creative ways of making housing more affordable. Churches should actually be excellent places for undertaking such thinking – they are often multi-generational communities that include those with substantial capital, and those with very little. If only there were some more creative and Kingdom-minded accountants in our midst! But the challenge is there to start thinking creatively and looking for examples of others who have already done so.[6]

In conclusion, I hope it is obvious, but it is worth stressing, that rejecting the spirit of the city does not mean rejecting the city itself or the genuinely good things that the city may offer. However it does require exercising a far more active, critical and rigorous discernment of the human-contructed environment that surrounds us, and it requires daily acts of resistance to those forces which dis-connect us from God, from each other and from the earth. It is central to the vocation of Christian communities to be people who live well in the midst of bad living – to be people who in the myriad of their daily activities choose for shalom, even in the heart of Babylon .

6. For one example, see Alison Sampson, "Investing in homes and people" in *Manna Matters*, December 2014 <www.mannagum.org.au/manna_matters/december-2014/everyday_people>

SO SHALL WE REAP

A biblical perspective on the agricultural challenges of the coming century

'Be not deceived; God is not mocked:
for whatsoever a man soweth, that shall he also reap.'
Galatians 6:7 (KJV)

In 2008 I sat in a rice field in Laos and listened to a proud farmer explain how he had increased his rice production and abandoned any need of risky credit. Xienloua, then 48, was one of the first in his village to pioneer an organic, low-cost, low technology form of rice production – the System of Rice Intensification (SRI) – first developed in Madagascar. Using this method, which he had learnt through an Oxfam project, Xienloua and his family of eight had increased the yield of rice from their single hectare of land, from around two tonnes to six tonnes and no longer needed to borrow at high interest rates to buy seed and fertiliser. A fantastic result for them, and now for their village, who are starting to emulate their efforts.

A few months later I sat in a ute in the New South Wales' Riverina district, and watched as an elderly rice farmer – a distant relative of mine – radioed in a helicopter to come and spray his 900 hectares of paddy with pesticide. He had shown me his shed with four tractors bigger than I had realised existed, and his irrigation pumps, which looked to me more like a metropolitan water-supply system. When I asked him about yields he casually replied that he was getting around ten tonnes per hectare. My head spun.

A Lao family of eight achieving, by a Herculean effort, the hitherto unheard of result of six tonnes from their measly one-hectare plot, and a single Aussie octogenarian casually extracting ten tonnes per hectare from his 900 hectares of land. These are two radically different agricultural worlds; but can they both continue to exist in the future that lies before us?

That same year, while we fixated on the GFC, another global crisis was unfolding whose human impact put the financial crisis in the shade. During the 2008 Food Crisis, the price of the three grains that supplies half of the world's calories – wheat, corn and rice – spiked dramatically, even though it was little noticed here. The sudden un-attainability of food for millions around the world sparked food riots in over thirty countries across three continents. Some see it as the spark that lit the Arab Spring. It is estimated that an additional 76 million people starved that year. For those watching the deep structures of the world economy, it seemed a harbinger of things to come.

In another thirty-five years the world will need to feed at least nine billion human mouths, and for many this simple fact determines what sort of agriculture will be needed. The weight of the numbers seems to push aside all other considerations, but do the numbers in fact conceal the bigger truth?

~

It is an indictment of the thinness of Christianity in the modern era that many will be surprised by the contention that the Bible has something vital to say about agriculture. Yet, not only does it have something vital say about agriculture, I believe it offers a prescient and historically accurate explanation of the crisis confronting humanity at the beginning of the twenty-first century. More than that, the Biblical vision of the interrelated subjects of human relations to the natural ecology, of agricultural production, and the form of our economic relations, offers, I believe the most realistic and most hopeful way for us to negotiate the coming crisis.

Crisis? If you look out your window you won't see people screaming in panic just yet, and indeed, if the content of our political debate is any indication, we are rather relaxed about our present position. However, many of those who by profession or interest spend time looking into the big picture of human affairs are getting decidedly anxious. Let me sketch out some of the components of the picture.

As mentioned above, by 2050, according to UN estimates, the human population will reach nine billion and plateau somewhere around that mark (some think the plateau will be at ten billion). It is expected that by this time, 70% of the world's population will live in cities (currently it is 50%), leaving perhaps 15-20% of people to raise enough food for everyone (currently it is about 30%). If current trends apply, there will also be less productive soil in which to raise food – over the last forty years perhaps as much as 30% of the world's arable soil has become unproductive, and by one reckoning, 10 million hectares of cropland is currently being lost annually to soil erosion. In many regions of the world, soil salinity and compaction are growing problems. On top of this, if climate change predictions are correct, then the timing and duration of rains upon which agriculture worldwide has been based, is likely to become seriously disrupted. Indeed, there is little doubt that this is already happening.

But not all of our food comes from the soil. Wild-capture fisheries (from both oceans and inland waterways) still supply a significant portion of humanity's protein intake – but not for long. Of the 232 ocean fisheries for which there is data, well over half have suffered a collapse at least 80% of their population, largely due to over-fishing, destructive fishing, and ocean pollution. Inland fisheries are in similar crisis; the world's most productive inland fishery – the Mekong River and its tributaries – currently supplies 70-80% of the protein consumed in that region. However, if currently plans for hydropower development proceed (as is happening) this fishery is also set to collapse dramatically, which means that all this protein will presumably have to be raised on land somehow, land which is becoming increasingly scarce.

I could go on, but you get the picture. More food is needed from less land and less people involved in growing food. This is widely accepted as the great challenge that is confronting us in the unfolding century, and it is undoubtedly unprecedented territory for humanity. For we who have suckled on the Enlightenment myth of progress, this is a shattering prognosis of the future. If only we had paid more attention to those texts that Christians call sacred, we would not be so surprised and confused.

For our current predicament is only what the Bible always said would happen if we followed the path that collectively we have chosen.

'Thus' says the Lord

Agriculture is the imposition of human culture upon nature; it is the imposition of order upon wildness. And culture, even secular culture, is always a development of cult – that is, our deep beliefs about the order of the universe. To uncover the depth of what the Bible has to say on agriculture we need to uncover what it has to say about the natural world, about the human place within it, and the human task in the world.

The Bible opens with a stunning hymn about the nature of nature and our place within it. While the dominant imperial creation myth of the time (the Babylonian *Enuma Elish*) taught that the world was born of the murderous and gory conflict of vindictive gods, Genesis 1 proclaims that the world is created out of a good God's intent – it was born of love – and that the fullness of this natural (created) order is good; indeed, it is *very good*. Ellen Davis, in her seminal book, *Scripture, Culture & Agriculture*, describes the Hebrew of Genesis 1 as almost falling over itself to emphasise the place of seed and fruit in creation. Where the ancient pagan fertility cults demanded that sacrifices be made to capricious Gods (and the kings who represent them) to ensure the next harvest, Genesis 1 proclaims that God has endowed creation with its own wondrous fecundity, a natural order that wills towards fertility. Where the *Enuma Elish* says that humans are created to be slaves of the Gods and their representative on earth (the Babylonian king), Genesis 1 states that every human being has an inestimable dignity – that of bearing the very likeness of God within themselves. Here, not in Ancient Greece, is the beginning of radical democracy. And, controversially, this image-bearing creature is given a special role within this wondrously fertile creation.

Contrary to the quite modern distortion that Genesis 1, in attributing 'dominion' to humans (vv.26, 28), has mandated the wanton exploitation of the earth, Davis demonstrates that the only possible reading that is faithful to the Hebrew text and its place within the narrative that follows, is that Genesis 1 is calling humans to the exact opposite of exploitation. She argues that for us moderns, a better translation is that humans are called to exhibit 'mastery among' the creatures of creation. Like a master craftsmen who works with reverence and respect for both his tools and materials, God's intention is that humans achieve a level of, reverence, respect, skill and understanding in working with creation that can be regarded as 'mastery'. And central to the idea of mastery, both for the Bible and the craftsmen, is the necessity of understanding and observing limits.

The second, complimentary creation story of Genesis (chapter 2) further establishes humanity's special role within creation, and thus the necessity of observing limits. In this story of human origins, *adam* (the human) is formed out of the *adamma* (the soil) – the human comes from the humus – and filled with the breath of God. This soil-creature, animated with the divine breath, is then given a foundational vocation 'to work and to keep' the garden in which he has been placed. The Hebrew words are stronger than our English renderings: the verb 'to work' (*abad*) is more fully 'to work for' or 'to serve', and is standardly used of the work of a servant for its master. The verb 'to keep' is the rich Hebrew word *shammar*, used to indicate careful nurture, as in the Aaronic Blessing: 'The Lord bless you and keep you'. But it is also used in the frequent exhortation to 'keep the commandments' of God. Such commandments are by definition limits on human conduct, and therefore we already see that here, care and nurture are intrinsically linked to observing limits. The same word is also sometimes translated as 'observe' ('Observe my Sabbath and keep it holy.') and here the English rendering well suits the subtlety of the Hebrew: to keep within limits requires that those limits be care-fully watched and understood. Thus, we might say that from the perspective of Genesis 2, the original vocation of humanity is 'to serve and observe' the earth.

The call for humanity to exhibit its mastery by the observing of limits is one of the primary themes of the unfolding instruction of the Hebrew Torah. The foundations are laid in the story of the manna in the wilderness. Here, in the archetypal story of salvation in the Old Testament, the Israelites are liberated from slavery to Pharoah's excess, where they undertook limitless work in his fields and building his store-cities. By contrast, in God's graced economy they are instructed to gather only enough for each day, to not store up it up, and to cease all economic activity one day a week.

The manna story sets the tenor for the laws of the Promised Land, which in their context, represent a detailed vision of a community of *shalom* – a community where people are in right relationship to God, to one another, and to the earth itself. The laws concerning agriculture form a significant component of this vision of shalom. The laws concerning harvesting and reaping, often referred to as the gleaning laws (Lev 19:9-10; Deut 24:20, 23:24-25), establish that while property rights are protected, they are not absolute; others, in this case the poor, also have a right to the fruit of the land. In effect, this means a farmer does not have the right to completely maximise productive efficiency; indeed, we might say that a level of 'care-full' *inefficiency* is built into the system. (Of course, I am using 'efficiency' in its very limited modern sense, which is completely focussed on *financial* efficiency – if we have a broader and more rational view of the proper use of people and resources, then it is not inefficient at all.)

Similarly, the Sabbath laws prohibit the maximising of labour efficiency. All people, whether slave or free, local or foreign are entitled a rest from labour; but not just the people, the labouring animals, too, are entitled to rest. The Sabbath laws extend to the land itself, mandating a rest for the soil every seven years (Lev 25:1-7). Once again, the right to maximise production is curtailed, this time with the intention of ensuring the continuing fertility of the soil for future generations. However, the command explicitly states that this is not just for human benefit, but also to benefit the 'wild animals of the land'; they also have their right to the fruit of the earth, and their proper place within the community to which humans owe a duty of care. Here is a clear view, which we are only just beginning to reclaim, that human agriculture does not exist as its own quarantined natural order – it always take place within, and remains dependent upon a far broader ecology, what Wendell Berry calls 'the Great Economy'.

In the great Jubilee vision of Leviticus 25 limits are also extended to the ownership and control of land. This remarkable set of laws, which sets out perennial re-distribution of alienated land back to its original heritors, is still much debated and argued over; however, what is clear is that it represents the pinnacle of a consistent Biblical vision of widely distributed access to land, and a horror of the consolidation of land in the hands of a few. In Micah's wonderful vision of a world set to rights, where shalom reigns, the final requirement is that 'each sit under their own vines and their own fig tree, and no one shall make them afraid' (Micah 4:4).

Obviously, in an ancient agrarian society, such a strong embrace of small-holdership represents a deep concern for social justice. But might it represent more than that? Wendell Berry, the pre-eminent agrarian prophet of our own age, has argued consistently that, even with modern technology, human 'mastery' of land – which requires deep and intimate knowledge of particular places – can only ever happen on a certain scale. Beyond that scale, the land will always pay a price for the breadth of human ambitions.

In summary, then, what we can see in the Old Testament law, is a fleshed-out picture of what the Genesis mandate of 'dominion' and 'working and keeping the land' might look like. And the picture we are given, again and again, is the need for humans to place limits on their own activities. What we *can* do and we *should* do are two different things. What is also clear is that the purpose of these limits is not to place a bit between our teeth and make us compliant rule-followers; the purpose is to ensure social, economic, ecological and spiritual health.

Masters of the universe?

You can see where this is going. In the age of science and modernity, humans have indeed come to consider themselves masters of the natural world. However, the modern view of mastery has been formed precisely around the conception of refusing to acknowledge limits, perhaps nowhere more so than in the field of agriculture. From the improvers and enclosures of the eighteenth and nineteenth centuries, through to the Green Revolution of the 1960s, and the current revolution in genetic modification, the increases in yields furnished by modern agricultural methods over the last two centuries have been nothing short of astounding. Beginning with the re-organisation of fields and selective breeding, moving to the adoption of mechanisation and large-scale control of water, through to the development of inorganic fertilisers and pesticides, then the application of advanced genetics to develop high-yielding varieties, and finally in the bursting of the limits of the gene itself, humanity has appeared at each step to transcend the known limits of nature.

Agriculture is no longer just agriculture, it requires new appellations: *scientific* agriculture; *intensive* agriculture; *industrial* agriculture. This last term is perhaps the most useful for describing the shift that has taken place: agriculture has undergone a wholesale shift from the craft of husbanding living things, to the commercial science of inputs, outputs, distribution, efficiency, markets and profit margins. As E.F. Schumacher notes, the ideal of industry is the elimination of living substances. The shift has, of course, been one of material organisation, but it has also been more profoundly a mental shift in how we think of food, the earth, and even of human community.

Now, a country like Australia can produce a surplus of grain, meat and dairy, with under 2% of the population involved in farming. The hubris is palpable. But at what cost?

The objection to industrial agriculture is not one of nostalgia. In a little remembered book from the 1950s, *Topsoil and Civilisation*, Tom Dale and Vernon Carter wrote:

> Civilised man was nearly always able to become master of his environment *temporarily*. His chief troubles came from his delusions that his temporary mastership was permanent. He thought of himself as 'master of the world', while failing to understand fully the laws of nature. ... One man has given a brief outline of history by saying that "civilised man has marched across the face of the earth and left a desert in his footprints".[1]

---

1. Quoted in EF Schumacher, *Small is Beautiful: Economics as if people mattered*, Abacus, 1973, p.84.

The image of 'a desert in his footprints' is perhaps the hyperbole of poetic license; nonetheless, it well describes what has perhaps been the central effect of industrial agriculture all over the world: the rendering of soil into dirt. Adam, who is soil brought to life, is himself taking the life from the *adamma*. In some regions of the world, 'desertification' in its literal sense is becoming the defining crisis.

But human agriculture is also making a desert of wild spaces. To earth-systems scientists, the current 'mass extinction event' – with species loss at perhaps 1,000 times the background level – warrants its own geological time period: the age of the anthropocene. And of all of humanity's impacts on the other creatures who need this planet, the clearing of habitat for agriculture is the single greatest.

The image of deserts can also be applied to the social landscapes being left in the footprint of industrial agriculture. While having 2% of the population producing a nation's food may seem like a triumph to technocrats, for rural communities and farming families it has been a catastrophe. Isaiah pronounces his horror against an economic system where house is joined to house, and field is added to field until 'you are left alone in the land' (Isa 5:8). However, where Isaiah was denouncing the predation of the rich, in industrial agriculture the joining of field to field has been *a condition of survival*. How many farmers have bought out their neighbours, not in triumph, but in tears? The economic decline of rural towns, the drift of children away from the farm, the increased scale and complexity of farm economies and the suicide rates of farmers, can all be plotted on the same graph. Berry, a farmer himself, writes with angry passion on this issue:

> It ought to be obvious that in order to have sustainable agriculture, you have got to make sustainable the lives and the livelihoods of the people who do the work. The land cannot thrive if the people who are its users and caretakers do not thrive.[2]

2. Wendell Berry, *Bringing It to the Table: On Farming and Food*, Counterpoint, 2009, p.16.

Industrial agriculture is rendering the soil and the human soul barren through its failure to acknowledge the limits of health and the conditions of goodness. The place in which we now find ourselves, ecologically and sociologically, has borne out strange warnings made by those ancient Hebrews:

> Your wrongdoing has upset nature's order
> and your sins have kept away her bounty. (Jer 5:26)

Again and again the biblical story draws a link between human sin and the ecological productivity of the earth. From the very outset of humanity's fall, the effect is felt in creation: Adam is told 'cursed is the ground because of you' and Cain learns that the earth soaked with his brother's blood 'will no longer yield its fruit to you'. The promised abundance of the Promised Land is always conditional - '*if* you follow the way that I am showing you' – and accompanied by dire warnings: if you go your own way and go after your own gods 'there will be no rain and the land will yield no fruit' (Deut 11:17). Leviticus 26 makes the chilling observation that if you do not give the land its Sabbaths, then it will indeed have them, but they will be Sabbaths of desolation (Lev 26:34).

What we so long assumed was the overly superstitious mindset of ancient peoples is now the single great fact with which we are confronted: human sin – our greed, our violence, our overestimation of our capabilities, our unwillingness to learn – has brought us to a place where we can no longer rely on the planet to yield its fruit and the rains to fall in season. We can now empirically verify that, mixing ancient and modern concepts, the curse of the earth is *anthropogenic*.

Feeding the world

To some minds, all this is evading the point. All this concern about soil, species and suicide is very nice, but the big question is:how are we going to feed 9-10 billion people? The proponents of the current system will argue that, whatever your objections to industrial agriculture (which they are wont to bracket as nostalgia, Ludditism, or hippy radicalism), there simply is no alternative for feeding the world. The numbers are wielded as a moral bludgeon: if you are against industrial agriculture, then you must be willing to let millions starve. But, of course, numbers lie.

There are a number of powerful misconceptions about the role of industrial agriculture in feeding the world. Perhaps most powerful is the argument that only industrial agriculture produces food in quantities that can keep the masses of the world's poor from starving. Norman Borlaug, the 'father' of the Green Revolution in the 1960s is credited by the proponents of industrial agriculture with saving a billion people from starvation. Yet while the Green Revolution undoubtedly dramatically increased yields, especially in Southeast Asia, it also by necessity created more thoroughly commercialised and globalised agriculture which led by turns to increasing farmer indebtedness, landlessness and concentration of land in the hands of a few. Millions were transformed from producers of food to consumers of food and left no choice but to try their luck in the merciless megalopolis.

The Western caricature of poverty in the developing world has so often centred on the 'backwardness' of agriculture, where lack of modernity is almost the definition of poverty. The nature and causes of poverty is a subject too deep and complex to explore here, suffice to say that a fair case can be made to suggest that perhaps the reverse of this caricature is more true. The most grinding poverty, the most desperate suffering, the most soul-destroying powerlessness of the modern age has not been experienced by those in 'backward' agriculture, but by those who have been dispossessed of their land and heritage by the juggernaut of progress; and the vanguard of that juggernaut has been 'agricultural development'. This was the case in the English 'agrarian revolution' in the eighteenth century and it is the case in Colombia, Nigeria, Bangladesh, the Philippines and Laos today.

And where can the dispossessed go but to cities? Was it coincidence or an omen that 2008, the year of the Food Crisis, was also the year that humanity became an urban species, with over half of us now living in cities? As Berry notes, the worst social and economic consequences of industrial agriculture have not been noticed, because they are 'erroneously called "urban problems"'.

Industrial agriculture is not a technique, it is a food system. Controlling the system are the multinational agri-corporations – Monsanto, Cargill, Bayer, Du Pont – who have an absolute strangle-hold on seed, on patent rights and on fertilisers and pesticides. They are partnered by the global food giants who own most of the brands we eat – Nestle, Kraft, Unilever, Coca Cola – and by the supermarkets who now account for nearly all of our food purchases. It is a system that dictates to farmers how to farm, that dispossess the vulnerable and turns them into dependent food consumers of a globalised food market. Its proponents – who are amongst the wealthiest and most powerful corporations on the planet – argue that it works well to produce an abundance of food, but this is a distortion on two serious counts.

Firstly, the global food system is more than abundant – it is *glutted*. It has long been known that around one third of food produced now goes to waste. A recent study in the UK is now arguing that the true figure might be closer to 50%. Whatever the number, anyone operating at any point in the food industry – from farming, to transport, to storage, to retail, to the home consumer – can tell you that the waste is colossal. All the talk of agribusiness about the challenges of producing enough for 9 billion people has been evading the scandalous fact that we probably already produce enough for 9 billion people. But that is not the biggest scandal.

In 2008, when so many more of the world's poor went hungry, there was never actually a shortage of grain. The causes of the Food Crisis were multiple and complex – they involved such things as oil prices, diversion of grain to biofuel and livestock and financial speculation on commodity markets – but it was fundamentally a *market-driven crisis* and not an agricultural crisis. That year the world had its largest wheat crop ever, and at the end of the buying season in the US there was a record amount wheat left-over in the silos. We live in a perverse system that simultaneously produces obscene excess and unconscionable hunger. Never before have the biblical prophets seemed so sane and reasonable.

~

This may all seem thoroughly depressing, but from a biblical perspective having a proper grasp on the bad news is the only way to properly comprehend the good news. And the first, most important piece of good news is that the earth, despite our depredations of her, remains yet fruitful enough to supply our needs, if only we can become gentler and fairer.

It is not within the scope of this essay to properly describe the movement of hopeful agricultural alternatives, but the contours are simple. There are now in most regions of the world farming practices which are productive yet 'non-industrial', making good use of our best scientific knowledge and the best of traditional insight, and which can be all characterised as methods which farm the soil first, and then the crop. 'Organic farming' is the best known of these methods, however it is too simplistic and puritanical a term to properly describe the range of farming practices which now seek to be kind to the soil and surrounding ecology. Another common factor in these movements is a return to farming on a smaller scale – involving more human labour on smaller pieces of land. By extension, a good case can be made that our best hope lies in reversing the great movement that most see as a *fait accompli* – the continuing urbanisation of the human species. No social engineering can accomplish this; however, what no government can achieve may yet be forced by what is going to happen to food prices. And accompanying such movements is both the need and the possibility of achieving some measure of re-localisation of food markets. Autarchy is neither possible nor desirable – the 100 mile diet is a fun exercise but not a broadly viable option – but some measure of regional food sovereignty is required to wrest power from the global behemoths who currently control the system.

The hope for humanity remains where it has always lain:

> If you follow my statutes and keep my commandments and observe them faithfully, I will give you your rains in their season, and the land shall yield its produce, and the trees of the field shall yield their fruit. Your threshing shall overtake the vintage, and the vintage shall overtake the sowing; you shall eat your bread to the full, and live securely in your land. (Lev 26:3-5)

As the Apostle Paul says, the calling of God is irrevocable. We are still people who are called to become masters, not of the natural order, but of ourselves: people who understand the limits within which we operate, most important of which is the limit of our understanding. The earth remains good, indeed very good, if only we can serve and observe her. The laws of creation are immutable: 'God is not mocked, for you reap whatever you sow ... So let us not grow weary in doing what is right, for we will reap at harvest time, if we do not give up.' (Gal 6:7-9)

THE LAW OF LOVE
Jesus, Paul and the ethics of eating

I abhor almost everything about multi-national fast food chains and what they represent. I abhor the sort of agriculture that is required to supply them and what this does to farmers, to animals, to the land and to the poor; I abhor the form of food that is produced and the resulting health impacts in our community, especially amongst the marginalised; I abhor the forms of marketing, advertising and branding they employ, and the ways in which it manipulates and distorts desire, family, sexuality, childhood and adolescence; and I abhor the style of business they represent, particularly how they drive out locally-rooted independent small-scale businesses.

But *geez* I love that chicken and those chips!

Every summer, when the cricket is on, saturated with fast food adverts and branding, I am plunged into a titanic spiritual struggle. Jesus got it right when he said that the spirit is willing but the flesh is weak – especially when the flesh is coated in those secret herbs and spices and a truckload of salt and MSG! As the cricket season progresses the tension becomes unbearable, and I inevitably end up making a surreptitious trip down to the local outpost of multinational fast food to buy some of that infernal fried and salty stuff, feeling more self-conscious than if I was buying pornography. To make matters worse, my good friend Nick Ray, author of *The Guide to Ethical Supermarket Shopping* and generally inspirational human being, lives just across the road. Damn!! Perhaps the Apostle Paul was fighting the aroma of those chicken fryers when he wrote: 'I do not understand my own actions. For I do not do what I want, but I do the very thing I hate. [...] Wretched man that I am!' (Rom 7:15-24)

What I am describing here is an example of the tormenting struggle between conscience and desire which many people would experience in one form or another. In this case, however, my struggle is entirely the product of my subscription to a self-imposed code of conduct that might loosely be called 'ethical consumption' or 'responsible consumption'. But there are many who quite understandably ask: is all that torment worth it?

Is it worth it, indeed? There are some real moral, theological and practical tensions inherent to the world of ethical consumption which are rarely articulated, even though, in my experience, most people who begin to explore this world struggle with these tensions in some way. Quite apart from questions around 'effectiveness', which loom large for many people, [1] even once the ideas of ethical consumption have been embraced, there remain some deeper questions about what is being pursued and why.

The idea of ethical consumption is founded on two simple primary principles:

1. Our need to reduce unnecessary and frivolous consumption, thereby reducing the strain on the earth's resources and on the other creatures who share this planet.
2. Our need to encourage production processes that take better care of people and the earth. Generally, but not always, this involves being prepared to pay a higher price.

However, as simple as the principles sound, actualising them in day-to-day life is immensely complex – the genius of our consumer system is that the true story of the impacts on people and places is entirely hidden from our view. That is why, over the last couple of decades, a huge amount of work has gone into developing some easily recognisable proxies for these principles that allow the average person to translate ethical aspirations into action at the checkout. These proxies are starting to become well known: Fair Trade, organic, No Sweat, free range, palm oil free, GMO free, 100% recycled etc. But there are

---

1. Does ethical consumption lead to an overall positive or negative impact on the poor? This seems like a simple question, but it is actually fiendishly complex, and much depends on the sub-questions one pursues to try and answer it. I have written elsewhere about some of these complexities – see J. Cornford, 'Unintended Consequences?: Ethical consumption and low-income labour', *Manna Matters*, August 2013, <www.mannagum.org.au/manna_matters/august-2013/understanding_the_times>.

other considerations too: company ownership, the amount and type of packaging, and transport miles.

In 2007 the Ethical Consumer Group produced *The Guide to Ethical Supermarket Shopping* that comes out in a new edition every year. The fact that *the Guide* sells more than 20,000 a year is an indication that concern about the impact of our consumption is not limited to a few fringe hippies and radicals. The *Guide* offers a simplified means of choosing between similar or identical products by distilling a huge amount of information about the record of the companies behind the products down to four different types of ticks or crosses. As I mentioned, Nick Ray, one of the authors of the guide is a good friend. Nick is painfully aware that such ticks and crosses cannot adequately represent the situational and moral complexity of the choices we are faced with; however, he is also painfully aware of the need to help people move beyond analysis paralysis. Thus, when standing at the supermarket shelf for olive oil, rather than agonise over a series of conundrums and lack of information, I can choose the one that is made in Australia, and owned by an Australian company that gets a tick for company record.

So I choose the products that get the tick, or have the Fair Trade badge, or are certified organic. And I try, despite myself, not to choose multinational fast food. But in following these proxies to guide what I buy, have I unwittingly subscribed to a new kosher? Do we now, in ethical consumption, have a new form of clean and unclean foods, the consumption of which marks the righteous from the unrighteous? If I say that I make these choices based on faith and conscience, am I saying that God requires them? Is not this then justification by works rather than faith? In short, is there a danger that by adopting an ethical code of conduct about what we buy and eat, we are in fact setting up a new legalism, the sort of religious system that was overthrown by Jesus and Paul?

To explore these questions requires untying, or at least loosening, some deep-seated theological knots: our attitudes to and understanding of the Old Testament law; our understanding of where Jesus, and then Paul, stood in relation to this law; and how this informs our approach to modern codes of conduct.

Torah re-visited

To usefully compare modern ethical codes of conduct to the Old Testament law – the *Torah* – we need to gain a fuller sense of both the positive and negative implications of such a comparison. Rather than 'the law', a more sympathetic interpretation of the word 'Torah' is *instruction* or *teaching*. Although, because of the huge gap of context it is hard for us to see, Torah is far from an arbitrary list of rules. It is, rather, a detailed, wide-ranging, holistic, integrated vision of what it would look like for humans to live in *shalom* (right relationship) with each other, with creation and with God. It addresses not just religious rules, but economics, politics, ecology and situational ethics. Torah not only provides a series of instructions and guidances on how individuals can conduct themselves ethically in the day to day complexities of life, but articulates a structure of society in which – as Peter Maurin of the Catholic Worker movement would have said – it is easier for people to be good. More than that, it is through the Israelites' obedience to Torah that they are to embody the character of God in the world. It is by living out this instruction in God's way of life that God's people are to tell the world about God

Both Jesus and Paul affirm the fundamental goodness of the intent of Torah. In the Sermon on the Mount, Jesus famously states: 'Do not think that I have come to abolish the law or the prophets; I have come not to abolish but to fulfill.' (Matt 5:17) Paul, in his extended discourse on the law in the letter to the Romans, declares that 'the Torah is holy, and the commandment is holy and just and good' (Rom 7:12).

Nevertheless, one of the defining conflicts within the gospels is between Jesus and those who have most staked their faith to Torah-observance, the Pharisees. Jesus' critique of the Pharisees is strident and unrelenting, pointing out that in their ever-more intricate development of rules to live by, the Pharisees have 'strained out a gnat but swallowed a camel'.

In Matthew chapter 23, Jesus pronounces an extended indictment of the rules-based religion of the Pharisees. The Torah that was intended to give guidance in the ways of justice and shalom has ended up squeezing out the place of love for one's neighbour, it has replaced the need for honest and humble self-reflection in the presence of God, and it has ultimately become a vehicle of death rather than life. Jesus' re-interpretation of a series of Torah commandments in Matthew 5, and his general unconcern for rigid Sabbath-observance (see Matt 12:1-8) reveal his purpose both to reclaim the intent of Torah, but ultimately to go well beyond it in fully revealing the way that leads to life.

And it is Paul, the once Pharisee, who, after his conversion encounter with the risen Jesus, is led to dramatically declare that those who are 'in Christ' are no longer under the law. Paul's life is gripped by the breathtaking insight that God's covenant with Abraham ('all peoples on earth will be blessed through you' - Gen 12:3) and the intentions of Torah (the faithful embodiment of the character of God) are all accomplished in Jesus. Paul understands that the whole meaning and intent of Torah has now come to fruition, which means it has taken new shape. And for Paul, the new shape of following God is summed up in one little phrase with a big meaning: 'faith in Christ'. Scholars such as NT Wright and Luke Timothy Johnson have argued that our English rendering of this pivotal Greek phrase (*pistis Christou*) does not quite do it justice. A fuller rendering would be something like 'faith in Christ' *plus* 'the faith of Christ'. Following rules and commandments might be an easily comprehensible way of practising religion, but it fails to achieve the profound transformation (the second birth) that God desires for us. For Paul there is now only one defining teaching and instruction, one 'Torah', to live by, and that is the person and the lived example of Jesus.

Paul is therefore horrified at the scurrilous suggestions that one can have an intellectual and abstract 'faith in Jesus' that then somehow allows one to ignore frameworks for living in right relationship. Paul is on the one hand adamant that the life of faith is the life of grace and therefore cannot be lived by a written set of rules; but on the other hand he is also adamant that the life of faith in Jesus requires the conforming of our whole conduct in this world to '*the law of the spirit of life in Christ Jesus*' (Rom 8:2). He calls for our bodies to be given as 'living sacrifices' (Rom 12:1) and declares that the ethical standard of life is now fundamentally simple, yet profoundly demanding:

> For you were called to freedom, brothers and sisters; only do not use your freedom as an opportunity for self-indulgence, but through love become slaves to one another. For the whole law is summed up in a single commandment, 'You shall love your neighbour as yourself.' (Gal 5:13-14)

Love does no wrong to a neighbour; therefore, love is the fulfilling of the law. (Rom 13:1)

Paul & ethical eating

Like Jesus, Paul's call is both liberating and daunting. What does it mean to love my neighbour in all the complex interactions of life? If we had no more guidance than this we might struggle to agree on how to interpret Paul's intention, but luckily we have a couple of instances where Paul works this principle through in relation to the ethics of eating, and he shows how his approach is finely nuanced to the complexities of situation and circumstance.

In 1 Corinthians chapters 8-10 and Romans 14 Paul addresses questions of conscience that have come up around eating in these two communities. In Corinth, a community with largely ex-pagan converts, a dispute has arisen as to whether Christians should or should not eat meat that has been sacrificed to pagan idols. In Rome, perhaps a more mixed community of Jewish Christians and Gentile Christians, Paul gives guidance on how these two entirely different food cultures can co-exist within one body. What is immediately striking when reading these passages is that Paul's guidance is not simple. Paul steadfastly refuses to lay down a rule about 'what is right', but rather insists his readers dig below the surface of their own ideas about food and pay attention to the *relational implications* of their actions. How do their decisions about food affect others?

In Corinth, it seems that some Christians, self-confident in their belief that there is only one God, have insisted that there is no harm in eating meat that has been sacrificed to idols, as such idols are not real. (Most butchery in the Hellenistic world was associated with the rituals of a pagan temple of some sort.) Paul agrees with them. Meanwhile, others in the community are not able to disassociate eating such meat from supporting the idolatrous religion that they have turned their backs on. Paul is entirely sympathetic with their position. What are they to do? Paul refuses to admit an absolute right or wrong with either partaking or abstaining, but rather insists on one principle:

> 'All things are lawful,' but not all things are beneficial. 'All things are lawful,' but not all things build up. Do not seek your own advantage, but that of the other. (1 Cor 10:23-24)

In particular Paul insists that 'the strong' (and surely his usage of this term is laden with some irony), those who are self-confident in their beliefs, show regard to 'the weak' (those whose consciences are fragile) and be prepared to change their eating habits for their benefit: 'if food can be the cause of a brother's downfall, I will never eat meat any more' (1 Cor 8:13).

In his letter to the Romans, Paul similarly refuses to take sides in their differences around eating, but points them to the same principle. While Paul recognises that there are different perspectives on the ethics of eating within the community, he is sharply critical of anyone whose adherence to one perspective has led them to become judgemental of those who differ. Rather than try and bring these groups to a common perspective on the Jewish food taboos (either for or against them), Paul is concerned that each person act with integrity to their own conscience: 'The faith that you have, have as your own conviction before God. Blessed are those who have no reason to condemn themselves because of what they approve' (14:22). However, Paul is also fundamentally concerned that each person's conduct take into account the good of the whole community: 'If your brother or sister is being injured by what you eat, you are no longer walking in love … Everything is indeed clean, but it is wrong for you to make others fall by what you eat' (v.15, 20). While Paul is not disagreeing with anyone's intellectual conviction, *in practice* he is asking that those who have no inhibitions about food and drink to nevertheless be prepared to accept some restrictions, for the sake of their brothers and sisters. What is crystal clear to Paul is that personal gratification should never get in the way of relationship: 'For the kingdom of God is not eating and drinking but justice, peace and joy' (v.17).

A modern Torah?

So what does all this mean for us now? Would Paul support contemporary efforts at ethical consumption, or would he see it as a barrier to 'the law of the spirit of life'?

What should be immediately clear from the above discussion, but what nevertheless still needs to be stressed, is that Paul is not at all interested in what we might call 'purity'. He shows absolutely no concern that what you eat or drink might somehow put you on the wrong side of God.

From my observation, there is sometimes a real danger that discussion of ethical consumption amongst Christians can implicitly assume – without ever quite articulating it – that the goal is 'not doing the wrong thing'; or to put it more bluntly, staying clean. Perhaps, even more worrying, the goal can even subtly shift to *being seen* to do 'the right thing'.

When ethical consumption becomes a code for 'clean' and 'unclean', then it must be rejected. For one thing, it would require all those proxies we have developed to guide ethical consumption to always be 'right' all the time (an impossible ask), or else the whole exercise becomes futile. Moreover, the idea that in this mind-bogglingly complex global economy we could somehow achieve a status of being 'pure', no longer implicated in wrongs of the world, is delusional.

But more seriously, as both Paul and Jesus understood, purity codes have the effect of creating division between people – of delineating those who are 'in' and those who are 'out', and further leading those who are 'in' to become judgemental of those who are not. And that is one thing that Jesus and Paul won't countenance: 'Who are you to pass judgement on the servant of another?' (Rom 14:4); 'Judge not, so that you may not be judged' (Matt 7:1).

More than once I have heard new converts to ethical consumption agonise over whether they should or should not drink the coffee at their friends' house, knowing that it is not Fair Trade. From a Pauline perspective, this is a non-issue: drinking a cup of Nescafe (that your friend has already bought) is not going to hurt anyone, however, refusing the hospitality of a friend (or anyone for that matter) has more serious relational implications. In our household we have made a decision not to buy any Nestle products because of their woeful corporate record, but it would be rude, ungrateful and plain wasteful not to accept and enjoy a box of Nestle chocolates that someone, acting out of kindness, has bought for us. The great spiritual danger of purity codes is that they become a substitute for, or even a barrier to, *faith*, that small-but-huge word that Paul uses to describe the ongoing process by which humans struggle to be oriented to the God of love.

So a concern for purity – something that supposedly keeps us on the right side of God – is not a reason that Paul would endorse for exploring ethical consumption; however, there are some much more substantive reasons to take up an ethical code of conduct in consumption, and these align closely with Paul's primary concerns.

As noted above, foundational to Paul's instructions on eating is the *relational implications* of people's decisions: 'For the kingdom of God is not eating and drinking but justice, peace and joy' (Rom 14:17). In this quote Paul is drawing on the big Hebrew concepts of justice/righteousness and peace/shalom (right relationship) that fill all his writings. It represents his conviction that through the coming of Jesus, God is undertaking the work of putting the world to rights – of establishing right relationship between people, between people and God, and between people and creation – and that those who are 'in

Christ' are called to participate in this great shalom-making purpose (see 2 Cor 5:17-20).

One of the great accomplishments of people such as Nick Ray, the Ethical Consumer Group and others like them, has been to lift the veil on the consumer economy and show how, through our acts of consumption, *we are in relationship* with people all over the world, and with the earth itself. And the reason this incredibly dense web of relationships is so ingeniously hidden from our view is that so much of it is exploitative and alienating, the opposite of justice and shalom. Through the frameworks of ethical consumption, however, we can, *acting out of love and from our own free will*, choose to restrict our own consumption and limit our own gratification in order to make the best choice that we can for the sake of our neighbour, and for the sake of God's good earth upon which we all depend. Surely this is an idea of which Paul would thoroughly approve.

When acting from this basis, we are acting according to what Paul calls 'the law [Torah] of the spirit of life'. Not only is it a choice of love, it is a choice of conscience, which is another way of saying it is a choice to integrate belief and action, and this also is critical for Paul. Knowing what we now know about our consumer system, how can we now read Jesus' challenging response to the question 'Who is my neighbour?' and continue to ignore the implications of our consumption for others? 'Blessed are those who have no reason to condemn themselves because of what they approve. But those who have doubts are condemned if they eat, because they do not act from faith' (Rom 14:22-23).

But this is exactly where we need teaching and guidance, because the complexity of the consumer system so effectively obscures what a choice for love might look like. The frameworks and proxies that have been developed around ethical consumption offer practical guidance - yes, a kind of Torah - for negotiating these complexities in our day-to-day choices. Indeed, by invoking the comparison to Torah, we very usefully gain a sense of the benefit, but also the dangers and limitations, of trying to live by such frameworks.

So let's embrace ethical consumption frameworks for what they are, and not imagine that they are something more. They are partial, contextual improvisations that help us to more easily make good choices in a global economy that is horribly broken and horrendously complex. They are not infallible and they are not the last word on what is right or good, and neither should we expect them to be. Tools such as *the Ethical Guide* are based upon the best information available, however, such information is never perfect or complete, and is changing rapidly. Certification codes such as Fair Trade and Certified Organic are systems which endeavour to guarantee better treatment of people and the land, however, all human systems are liable to break down somewhere along the line. Don't be dismayed or even surprised when some certification code is shown to be flawed in some way – they too will always need scrutiny, critique and improvement. Don't let our inability to make 'the perfect choice' (whatever that is) stop us from making the best choices that we have available to us. What the world needs of us and what God hopes for us is not that we attain moral perfection, but that we form habits in trying to choose what is good, acting out of love for our neighbour and for the earth, even if we sometimes fail, and even if we sometimes just can't quite resist slipping down to the local fast food chain …

The other day I read an article in a gardening magazine that rhetorically declared, 'What could be better than growing your own organic kale?' I reckon I could think of a few things. Top of my list would be if I could get my hands on some locally-sourced, organic, free range, Kentucky-style fried chicken made by a locally-owned, independent small business! I reckon I might just pass up the organic kale for some of that …

The Torah of Ethical Consumption
Seven guiding principles

1. Buy less stuff!

2. Choose longer-lasting and better quality.
- less plastic
- less toxic materials
- more durability
- less packaging
- more recyclability

3. Can you buy what you need second hand?
- clothing, technology (phones, computers, TVs etc), furniture and other household items, cars and bikes, and much more ...

4. Where possible, choose products with certification systems that provide some protection for people and the environment, in particular, look for:
- Fair Trade
- Organic and Free Range
- Ethical clothing Australia
- Palm Oil free – Roundtable on Sustainable Palm Oil
- Marine Stewardship Council certified seafood

5. Preference products marked 'Product of Australia'.
This means that the bulk of the materials and manufacturing have been sourced here; 'Made in Australia' refers only to manufacture.

6. Use *The Guide to Ethical Supermarket Shopping* to:
- preference Australian-owned companies
- preference companies with better corporate records
- avoid companies with poor corporate records

7. Agitate for change.
Support efforts to encourage the major brands and retailers to take more responsibility for workers' rights and the environment.

www.ingramcontent.com/pod-product-compliance
Lightning Source LLC
Chambersburg PA
CBHW051432090426
42737CB00014B/2928